THE CONCEPT OF SCIENTIFIC LAW IN THE PHILOSOPHY OF SCIENCE AND EPISTEMOLOGY

BOSTON STUDIES IN THE PHILOSOPHY OF SCIENCE

VOLUME 208

THE CONCEPT OF SCIENTIFIC LAW IN THE PHILOSOPHY OF SCIENCE AND EPISTEMOLOGY

A Study of Theoretical Reason

By

IGOR HANZEL

Comenius University, Slovak Republic

KLUWER ACADEMIC PUBLISHERS

DORDRECHT / BOSTON / LONDON

Library of Congress Cataloging-in-Publication Data.

A C.I.P Catalogue record for this book is available from the Library of Congress.

ISBN 978-90-481-5275-9

Published by Kluwer Academic Publishers BV,
P.O. Box 17, 3300 AA Dordrecht, The Netherlands.

Sold and distributed in North, Central and South America
by Kluwer Academic Publishers, PO Box 358,
Accord Station, Hingham, MA 02018–0358, U.S.A.

In all other countries, sold and distributed
by Kluwer Academic Publishers, Distribution Centre,
P.O. Box 322, 3300 AH Dordrecht, The Netherlands

Printed on acid-free paper

In memory of my parents

TABLE OF CONTENTS

PART 3: THE INHERENT LAW OF ESSENCE

INTRODUCTION

This book is primarily a study of the concepts of scientific law and scientific explanation. It is based on the idea that the natural and social sciences do not operate exclusively with one type of scientific law and that they do not explain phenomena by means of one exclusive method. Because I understand explanation based on scientific laws as one form of the appropriation in spirit (appropriation in mind) of the real (natural and social) world by human beings, this idea amounts to the claim that science is not mono-paradigmatic, but poly-paradigmatic. In the following chapters, I realize this idea through a reconstruction of three types of scientific law: the law of appearance, the pure idealizational law of essence, and the inherent law of essence, and through a reconstruction of the methods of explanation based on them. I view these three types of scientific law and methods of explanation as representatives of stages in the development of the appropriation in spirit of the real world. But because they are just stages of one process, their truth resides, so to speak, in their unity. Problems which can be formulated at one stage, in one paradigm, but cannot be solved in its framework, very often find a natural solution in one of the others.

Such a developmental approach to the issue of scientific law and explanation requires a special effort from the reader, who is asked to consider the following chapters in their unity. The discussion requires from the reader educated in the Anglo-Saxon philosophical tradition a willingness to pass from a certain paradigm on to the others, a willingness to loosen one's grasp on a single paradigm as the *ultima ratio* of the understanding of the world. The discussion requires the willingness to leave behind the standard understanding of the concept of essence as either an eternal Platonic form hovering over the real world or as the Kantian noumenon inaccessibly hidden behind the phenomenon. The reward for this will be, I hope, an enlargement of the stock of categories guiding the production of thoughts in natural and social science.

The radical departure which I have tried to pursue in this book as I reconsider the standard understanding of categories like phenomenon, essence, measure, conditions of a scientific law, etc., has relied on and been prompted by contemporary work in the field. One positive impetus for this book came from the recent development of philosophy of science in Europe, development most evident in the work of philosophers like Leszek Nowak from the Poznań School in Poland and in the work of my teacher, Václav Černík from Bratislava, located in former Czechoslovakia. Their work represents a starting point for a new stage in the development of philosophy of science, because, firstly, they have been able to approach a number of unresolved problems in the philosophy of science, for example the problem of the reconstruction of explanation of a scientific law from another scientific law, by explicitly drawing upon the idea of the existence of underlying structure hidden, so to say, behind the empirical phenomena. Secondly, they have been able to approach Karl Marx's economical works on the basis of this idea and reconstruct certain aspects of his understanding of the concepts of "scientific law," "scientific explanation" and "conditions of a scientific law." Regarding the latter, Marx's ability to differentiate between two types of conditions of scientific laws – modification conditions and essential (inherent) conditions – has proven to be significant and useful. I realize that it is highly unusual, to say the least, to appeal to Marx in an analysis of the nature of scientific laws and explanation, but it is my contention that by means of this typology it becomes possible to enlarge the conceptual framework of the philosophy of science.

Another positive impetus for this book came from the publications of works of Karl Marx in the second edition of the MARX-ENGELS GESAMTAUSGABE (MEGA^2), which were accessible neither to philosophers from Poznań nor to V. Černík. This compilation of Marx's work enabled me to investigate the analytic laboratory of Marx's thinking in a new way and thereby to arrive at an understanding of the categories mentioned above which differs both from standard interpretations as well as from the more recent attempts to revise these interpretations, e.g., in P. Murray (1990).

In chapter one of this book, I deal with Hempel's D-N model of scientific explanation and reconstruct the notion of a scientific law of appearance which is closely connected to it. In chapter two, I scrutinize the basis of the scientific law of appearance and, therefore, also, of the whole D-N model, by analyzing the now classical discussion between W. Kneale, whose argument represents the necessitarian approach, and K. R. Popper, whose argument represents the regularity (Humean) approach. By analyzing this discussion I show that the clue to a post-Humean solution of the unresolved problems with respect to the concept of scientific law lies in the introduction of the concept "essence of a natural kind."

The first type of scientific law, which already expresses the knowledge of essence of a natural kind, is presented in chapter three. This is the idealizational type of scientific law, which was initially reconstructed by the Poznań School and especially by L. Nowak (1971; 1980). Nowak arrived at his understanding of the concept of "idealizational law," and at his model of scientific explanation, by analyzing physical theories and Marx's *Capital*. But his analysis is incomplete, and therefore I introduce the concept positing a "pure idealizational type of a scientific law of essence," and, in connection with it, I propose a model of scientific explanation which corresponds to this type of scientific law. I then show that though this model satisfies the explanatory procedures of classical mechanics, it reconstructs neither the explanatory procedures from the first volume of Marx's *Capital* nor the explanatory procedures from the manuscript of its third volume. Through an analysis of these explanatory procedures, accompanied by an analysis of those works by Marx that laid the groundwork for them (*Grundrisse 1857/58, Manuscripts 1861/63; Manuscripts 1863/67*), I show that my proposed concept and model are based on richer explanatory procedures than explanation by gradual concretization, and that they are based on a type of scientific law different from the pure idealizational type of scientific law of essence. In connection with the latter, the "scientific law of essence," I reconstruct its basis in chapter four, which primarily consists of a discussion exploring the concepts "natural kind" and "natural necessity."

Part three of the book is devoted to the reconstruction of inherent types of scientific law. In chapter five, I analyze the notion of a "scientific inherent law of essence," a "scientific inherent organic law of essence," a "scientific developmental law of essence," and a "scientific historic law of essence." I show that the first volume of Marx's *Capital* is based upon such types of scientific law, and I also reconstruct the methods of scientific explanation derived from them.

In the modern philosophy of science, the concept of measure has not been fully explicated.[1] Its importance lies in the fact that it enables one to reconstruct the relations of the knowledge of phenomena to the knowledge of underlying structures hidden behind them. I offer an explanation of this category in chapter six and distinguish there its three types: apparent (external) measure, immanent measure, and manifestation of immanent measure. Through their explication I will show that the knowledge of the first is given in the pure idealizational type of a scientific law, the knowledge of the second in the inherent types of a scientific law, and the knowledge of the third in the inherent idealizational type of a scientific law.

Chapter seven is an attempt to create conceptual means for a philosophical analysis of non-linear dynamic theories of highly organized complex systems as they are constructed, e.g., in contemporary non-equilibrium non-linear thermodynamics. In this section, I compare the approaches to these systems based, first, on the more traditional assumption that the system being studied is externally regulated and, then, on the assumption that the system being studied is an open system with internal regulation. The latter, as compared to the former, represents a more advanced conceptual tool for the scientific investigation into complex systems, e.g., for the understanding of their self-organization and self-development.

Chapters eight to eleven deal with the ontological aspects of the scientific inherent law. In chapter eight I start with an analysis of the ontological dimensions of Kripke's essentialism, and then I determine the type of necessity given in the inherent types of a scientific law. Here I show that essence should be understood not only as something which is related to phenomena, but also as something inherently relational, as a relation between the constituents of the internally differentiated essence. The latter approach, in my view, enables one to revive and further develop an ontologico-epistemological construal of essentialism. Simultaneously it allows essentialism to develop its own conceptual apparatus independently from Kripke's semantics of modal logic.

Chapter nine analyzes the problem of causal loops as well as the type of universality granted in the inherent types of a scientific law. Through this discussion I introduce the categories of the substance-subject and of the substance-universal and show their importance for the analysis of scientific theories dealing with self-reproducing objects.

In part four of the book, I locate the philosophical investigation of scientific laws in a broader epistemological context. In chapter ten I outline the relation of ontology, epistemology, and logic, and then explain the concepts of logical category, philosophical category, and a system of philosophical categories. In connection with the latter, the concept of a reproduction cycle of cognition will be introduced. This concept, in my view, enables one to grasp the phases through which cognition gradually passes. I show that the formulation of scientific laws and scientific explanation constitutes but a moment in the reproduction cycle of cognition. I show also that the system of philosophical categories should be built in such a way that its "construction clusters" successively cover all the phases of this cycle.

Reconstruction of the categorial basis of Marx's *Capital* centers chapter eleven; this analysis enables one to compare the system of categories laid out in Hegel's *Science of Logic* with the system of categories that constitutes the foundation of *Capital,* as well as of the preparatory work leading to it (*Towards a Critique of Hegel's Philosophy of Right*; *Grundrisse 1857/58*; *Manuscripts 1861/63*; *Manuscripts 1863/67*). Till now any epistemological analysis of these works had to be measured by Hegel's *Science of Logic*, simply because we still lack an adequate realistic (materialistic) counterpart to it. By comparing Marx's economical works with Hegel's *Science of Logic* I reconstruct their main points of agreement and their principal disagreements. This should in turn, I hope, pave the way for the creation of such a realistic (materialistic) counterpart to Hegel's *Science of Logic*.

In the conclusion of the book, finally, I outline the differences between the methods applied by natural science and methods applied by social science. I show that the method of natural science can be characterized as an approach to the object under investigation from the outside, and that the conceptual apparatus of the theory of self-reproducing (organic) systems is sufficient for understanding the objects under investigation in such cases. This conceptual apparatus and that method are necessary, but not sufficient, in the case of investigation into social systems. What is required here is a unification of the approach to the object from the outside with a coordinated study of the object from the inside (with the view through the "eyes" of the actors involved) by means of the unification of the theory of self-reproducing systems with the theory of action.

In the book, for the sake of brevity, I use symbolic formulae to represent the structure of scientific laws and the structure of scientific explanation. They are consecutively numbered in such a way that the first digit gives the number of the chapter where the formula was introduced.

ACKNOWLEDGEMENTS

Several persons and institutions were helpful in the writing this book. My interest in the nature of scientific laws was first stimulated by my former teacher, Professor Václav Černík. Chapters 1, 3 and 5 were written at the University of Pittsburgh, were I spent the Fall 1992 term on an IREX (International Research and Exchange Board) stipend. Chapters 2, 4, 6, 7 and 8 were written back at Comenius University. Professor Černík and Professor František Gahér from Comenius, as well as Professor Jozef Viceník from the Philosophical Institute of the Slovak Academy of Science, read drafts of this work and suggested several improvements. Finally, Chapters 9 through 11 were written at SUNY-Cortland where I spent the academic year 1996/1997 as a Fulbright scholar. Professors Kathryn Russel and John Ryder from this institution read the whole manuscript and made valuable proposals to improve it.

I am especially grateful to Professor Roman T. Ciapalo from Loras College, Dubuque, Iowa, who over the years read all the drafts of the book and made numerous suggestions for improving them. Thanks also to Professor Robert S. Cohen who took an interest in the manuscript of the book; without his support it would never have been published.

I am grateful to the Springer Verlag for the permission to reproduce the schemes of hypercycles published in (Eigen-Schuster 1979) and which were first published in the journal *Die Naturwissenschaften*, issues 11/1977, 1/1978, and 7/1978. I thank also the editor of the journal *Metaphilosophy* for permission to use in Chapter 10 material from a previously published article (Hanzel and Černík and Viceník 1994). I am grateful to Professors Černík and Viceník for letting me to use this material from our article.

PART 1

THE LAW OF APPEARANCE

[I]f the "phenomena" are appearances of something else, then the mere fact that this "something else" is that particular reality of which that particular phenomenon is the appearance – this fact enables us to describe reality just as completely as the appearance of it. The description of the appearance *is*, at the same time, a description of that which appears. The phenomenon can be called an appearance of some reality only in so far as there is some correspondence between them, they must have the same multiplicity; to every diversity in the phenomenon there must be a corresponding diversity in the appearing things, otherwise the particular diversity would not form part of the phenomenon qua phenomenon, nothing would "appear" in it. But if this is so, it means that the "appearance" and the "appearing reality" have identically the same structure. The two could be different only in content, and as content cannot possibly occur in any description, we conclude that *everything* which can be asserted of the one, must be true for the other also. The distinction between appearance and reality collapses, there is no sense in it (Schlick 1979a, 359).

[T]he Vienna Circle declared any question (and any answers) of a transempirical sort to be factually meaningless (Feigl 1969, 5).

CHAPTER 1

THE LAW OF APPEARANCE

The intent of this chapter is to recover the grounds underlying the D-N model of scientific explanation by examining C. G. Hempel's writings (1948; 1952; 1958; 1964; 1965a; 1966; 1970; 1988). Together with the concept of "scientific law" that it incorporates, the model had been used for a long time as the standard according to which one can assess an actual developmental stage of a given branch of scientific knowledge and, therefore, identify whether the discipline under scrutiny has or has not reached the level of a true science. Now, as an example of a concrete science whose explanatory procedures conform to the D-N model and whose scientific laws are in line with the concept of scientific law in this model, one would commonly refer to classical mechanics. Classical mechanics, then, may be set forth as the pattern for other scientific disciplines to follow. I shall begin by briefly outlining the problems and paradoxes illustrative of the D-N model. Next, I will reconstruct the Humean conception of scientific laws as one of pivotal points of the D-N model tacitly employed by the latter. By analyzing some scientific laws and explanatory procedures contained within classical mechanics, I will expose the paradoxes and problems of the D-N model and demonstrate that these problems stem from a Humean construal of scientific laws.

1.1 THE D-N MODEL: ITS PROBLEMS AND PARADOXES

C. G. Hempel first defines potential explanation, maintaining that it may be regarded as a "potential answer to a question of the form 'why it is the case that p?', where the place of 'p' is occupied by an empirical sentence detailing the facts to be explained" (Hempel 1965a, 421); he then identifies a scientific explanation as a potential explanation with true premises. Let a sentence describing the facts explained be signified as *E*. According to the D-N model, then, the explanation assumes the form of a deductive argument (Hempel 1965a, 336; 1966, 50). Here the sentence *E* is deductively derivable from sentences $L_1, \ldots, L_k$ representing the general laws on which the explanation rests, as well as from sentences $C_1, \ldots, C_k$ representing certain particular facts – singular conditions. Because these laws cover the explained facts one can speak here about a covering-law explanation. The sentence *E*, according to Hempel, may be a description of an individual phenomenon which occurred in a certain time and place. This is the case where the scientific laws $L_1, \ldots, L_k$ are empirical ones. Yet *E*, Hempel suggests, may have the nature of an empirical law (characterized by general regularity). Thus the scientific laws from which the explanation derives are theoretical laws, theoretical principles, or empirical laws of a broader scope than the empirical law to be explained (Hempel and Oppenheim 1948, 247; Hempel 1965a, 343; 1966, 51).

The essential characteristics of the D-N model briefly surveyed here were expounded by Hempel in the first part of his study (1948) and were considered again in his later inquiries (1958; 1965a; 1966). But in his attempt to provide a definition for scientific explanation by means of a simple model language in Part III of his study (1948), Hempel has already encountered his first problem. While he insisted that the sentence *E* might be also a scientific law, the process of definition brought him to a recognition of the problem which he articulated in the celebrated footnote 33, a problem which forced him to restrict his reconstruction of scientific explanation to apply only to a case of explanation of singular events (Hempel and Oppenheim 1948, 273):

> This is not a matter of free choice: The precise rational reconstruction of explanation as applied to general regularities presents peculiar problems for which we can offer no solution at present. The core of the difficulty can be indicated briefly by reference to an example: Kepler's law, *K*, may be conjoined with Boyle's law, *B*, to make a stronger law, *K.B*; but derivation of *K* from the latter would not be considered as an explanation of the regularities stated in Kepler's laws; rather, it would be viewed as representing, in effect, a pointless "explanation" of Kepler's laws by themselves. The derivation of Kepler's laws from Newton's laws of motion and of gravitation, on the other hand, would be recognized as a genuine explanation in terms of more comprehensive regularities, or so-called higher-level laws. The problem therefore arises of setting up clear-cut criteria for the distinction of levels of explanation or for a comparison of generalized sentences as to their comprehensiveness. The establishment of adequate criteria for this purpose is as yet an open problem.

But, though Hempel would later (1958; 1965a; 1966) insist that *E* in a D-N explanation might be a scientific law, he never settled this problem. Thus, his contention has not been verified, and the question of how scientific laws are explained remains completely open within the D-N model.

As I see it, the difficulty described in footnote 33 is but a constituent of a more complicated issue expressible through the following paradox (Kyburg 1965). The question "Why has this lump of table salt, after being placed into warm water, dissolved?" can be responded to in this way: "If placed into warm water, all lumps of table salt subsequently dissolve" (the law *L*), with the conjoint introduction of the singular condition, "This lump of table salt has been placed into the vessel which contains warm water." Yet another explanation seems to be just as plausible and coextensive with the D-N model: "All lumps of table salt, if placed into warm water and on condition that simultaneously a magic spell is pronounced over them, will subsequently dissolve," and this, again, with the singular condition, "This lump of table salt has been placed into warm water and a magic spell has been pronounced over it." If in its structure real scientific explanation corresponded to the D-N model, it would be impossible to prevent irrelevancies (e.g., Boyle's law as irrelevant to Kepler's laws) from slipping in. Their inclusion into a deductive argument, as we have seen, does not affect its validity, yet on the other hand it has a destructive effect on scientific explanation. The D-N model is also beset by further paradoxes (for a complete list of them see Salmon 1990, 46–50); the pendulum, flagpole (Bromberger 1966), and barometer (Jeffrey 1969) paradoxes have been selected for consideration here.

In the pendulum paradox, from the equation $T = 2\pi\sqrt{l/g}$, where "$\sqrt{}$" stands for square root, for the simple pendulum, one can account for a particular period of time measuring the oscillation of a given pendulum, this period being determined by virtue of the singular condition which characterizes its length. But the concrete length of a pendulum can also be calculated from the equation $l = T^2g/4\pi^2$ in conjunction with the individual condition which characterizes the value of the pendulum's period.

A similar difficulty arises in the case of the flagpole paradox. Consider a flagpole of a certain height h' brightly illuminated by the sun, whose rays are falling on it at angle k'. The shadow cast by the pole has length l' and the general functional dependence assumes the form $l = l(h, k)$. In an answer to the query "Why does a certain flagpole cast a shadow of such-and-such length?" one will explain that this length is just the result of the application of the law involving the equation for l to concrete values h' and k' By reformulating the original equation, however, one will arrive at the functional dependence $h = h(l, k)$, which permits one to "explain" why the pole's height is such-and-such by virtue of the invoked equation conjointly with the introduction of the particular values k' and l'.

The barometer paradox rests on the following consideration. Let us have a law "Whenever and wherever at a specified time t a sharp drop in the reading of the barometer occurs, at a time t plus two hours it will be raining." In answer to the query "Why was it raining at this or that place at 12 o'clock?" we will respond by invoking the quoted law conjointly with the stipulation that "At such-and-such concrete place at 10 o'clock there occurred a drop in the reading of the barometer." Despite the fact that the explanation conforms to the D-N model, it provides no account of the *cause* of the rain. The latter's cause is here presented by a certain atmospheric process of a fall in pressure which affects two mutually independent outcomes: the movement of the barometer's pointer and the falling of the rain. We are lured into the barometer paradox by the neglect of the cause-effect relation in the scientific law, which is selected to provide a foundation for explanation. The same applies to the flagpole paradox. The former's length is set regardless of whether the sun's rays are falling or not, i.e., the flagpole's height is not an effect of their influence. Even this intuitive analysis of the paradoxes makes it readily clear that the construal of causality involved in the D-N model as it is embedded in Hempel's understanding of the concept "scientific law" must be emphasized and examined.

1.2 THE HUMEAN SCIENTIFIC LAW

In Hempel's writings, scientific laws constituting the foundation of the D-N explanation of particular phenomena are subjected to a detailed analysis. They are the scientific empirical laws which express a "uniform conjunction between different empirical phenomena or different aspects of one and the same phenomenon" (Hempel 1966, 54). Their logical structure amounts to, "Whenever and wherever there occur certain conditions of the type F, there always and without exception occur certain conditions of the other type G" (Hempel 1966, 54). So let me express the empirical law in this way:

$$L_e : (x)(Fx \rightarrow Gx) \tag{1.1}$$

As an example of such a law Hempel offers the statement "Whenever the temperature of a gas increases while its pressure remains constant, its volume increases" (1966, 54). The same structure, according to Hempel, can be found in the statement "Whenever a body falls freely from rest (i.e. it is released) in a vacuum next to the surface of the earth, the distance it covers in t seconds is $16t^2$ feet." These scientific laws, occurring in a D-N explanation, are referred to by Hempel as *causal laws* since they (Hempel 1966, 53)

are always presupposed by an explanatory statement to the effect that a particular event of a certain kind G (e.g., expansion of a gas under constant pressure; flow of current in a wire loop) was *caused* by an event of another kind F (e.g., heating of the gas; motion of the loop across the magnetic field). To see this, we need not enter into the complex ramifications of the notion of cause; it is sufficient to note that the general maxim, "Same cause, same effect," when applied to such explanatory statements, yields the implied claim that whenever an event of kind F occurs, it is accompanied by an event of kind G.

In Hempel's view a *scientific causal law*, as a variety of the law of the type L_e, delineates a recurring succession of phenomena of a certain type. It expresses the relation of cause and effect in such a way that the former is understood as an antecedent phenomenon, and the latter as a consequent phenomenon. The relationship can be schematically represented thus:

$$L_{cs} : (x)(Nx \,\&\, Cn_{1-k}x \,\&\, C^o x \rightarrow E^o x) \tag{1.2}$$

This is to say that "Whenever and wherever there are objects of a certain natural kind N and 1 to k conditions Cn and phenomena-events of type C^o occur, then always and without exception phenomena-events of type E^o occur." C^o represents the cause and E^o the effect. Scientific explanation of the D-N type involving scientific laws with a structure corresponding to (1.2) has been called a causal explanation by Hempel. Apart from this, Hempel also mentions the case of a *non-causal* D-N explanation (1965a, 352). He maintains that as such an explanation may qualify, e.g., explanations of empirical laws by reference to theoretical principles or empirical laws of a broader scope (e.g., inference of Galileo's law of free fall, of the law for the simple pendulum, and Kepler's laws from Newton's laws), as well as qualifying some explanations of individual phenomena from empirical laws. By way of an example of such a non-causal law, Hempel refers to the law for the simple pendulum containing the equation $T = 2\pi\sqrt{l/g}$. Since, according to Hempel, the length l and period T are given simultaneously, this law represents a *scientific law of coexistence*, and a D-N explanation of an individual value of the period is not a causal one. The structure of empirical laws of coexistence can be expressed as follows:

$$L_{cx} : (x)(Nx \,\&\, Cn_{1-k}x \,\&\, E_1 x \rightarrow E_2 x) \tag{1.3}$$

where E_1 and E_2 denote coexisting types of phenomena.

Let me now consider Hempel's construal of causation as given in L_{cs}. In terms of its statement, the proposition that would answer to the barometer paradox I mentioned earlier would be a law of such a type. The property of regularity as well as of temporal succession, as given in Hempel's construal of scientific causal laws, is virtually rooted in Hume's understanding of the concept of causation.[1] In conformity with Hume there obtain but three objective characteristics of the cause-effect relationship in terms of discrete, atomic events: their temporal and spatial contiguity; their temporal succession; and their regularity (i.e., recurrence of their temporal sequence). This constitutes a covert ontological foundation which Hempel never explicates in his treatment of the concept "scientific causal law."[2] By explicitly assuming that science's aim ought to be the pursuit of the discovery of empirical regularities in the "flux of events," on which basis this theory then frames empirical laws (Hempel and Oppenheim 1948, 250; Hempel 1958, 173), Hempel implicitly builds up his construal of the concept "scientific law," as well as the D-N model, on a Humean ontology. So, then, at this point, there obtains the following connection: a certain ontology gives rise to a certain construal of the concept "scientific law," which in turn engenders a certain model of scientific explanation. If, however, the D-N model slips into problems and paradoxes when confronted with certain explanatory procedures in physics, then I am led to the conclusion or hypothesis that laws of physics are formulated in such a way that their structure differs from that of the formulae (1.2) and (1.3). Hence, presumably, they are shaped on ontological grounds of causation different from Hume's. This hypothesis seems tenable for one more reason. The paradox of the pendulum has emerged out of the impossibility of distinguishing, within a Humean construal of scientific laws, between the equation $T = 2\pi\sqrt{l/g}$ and the equation $l = T^2 g/4\pi^2$. The science of mechanics, however, *respects the differentiation.* The oscillations with a certain period are generated by the force acting on the pendulum; they are its effects. Because Hempel's approach rests on a Humean determination of causation, there is no way to distinguish between the case when the equation of a scientific law expresses an *ontic bond* of the genesis of an effect from its cause and the case when the equation has been amended and expresses merely an *epistemic approach.* The procedure of amending the original equation allows us to trace back to the cause issuing from its effect. To verify this hypothesis it is necessary to find out whether scientific laws, e.g., those of classical mechanics, have a structure corresponding to the structure of the formula (1.2) or (1.3).

1.3 LAWS OF MECHANICS

Let me put down the logical structure of the scientific law for a simple pendulum, partially exploiting the mode of expression as it has been given by L. Nowak (1971; 1972; 1980). The equation $T - 2\pi\sqrt{l/g}$ pertains to pendulums which meet these eight idealizations: 1) the friction at the fulcrum equals zero; 2) the whole mass of the pendulum is contained in the body suspended; 3) non-gravitational forces are not acting; 4) the angle of deviation is from the interval of (0, 5) degrees, i.e., $\sin k = k$; 5) the gravitational force remains unchanged in time; 6) the length of the suspension cord is constant; 7) the volume of the body suspended equals zero, i.e., it is a mass point; 8) the resistance of the environment is zero, i.e., the pendulum's motion occurs in a vacuum. Let me schematize each idealization in the form *Id.* The antecedent of the scientific law of the simple pendulum involves a conjunction of eight idealizations Id_1 & . . . & Id_8. If I express this conjunction as Id_{1-8}, and $T^{(8)}$ is the period of oscillations of the pendulum (denoted by "P"), while all the aforementioned idealizations are valid, then the logical structure of the law of the simple pendulum will look like

$$L^{(8)} : (x)(Px \,\&\, Id_{1-8}x \rightarrow T^{(8)}x = 2\pi\sqrt{lx/g}) \tag{1.4}$$

This law has been formulated in the context of physical theories which take into account the existence of gravitational as well as non-gravitational forces. That is why the scientific law for the simple pendulum already mentions within this context the *cause* that brings about oscillations. It is the force F which acts upon the pendulum of length l, and which, therefore, comes out as the cause generating the period $T^{(8)}$ of the oscillations. So, the scientific law for the simple pendulum as expressed by formula (1.4), I would argue, is an instance of a *theoretical causal law*. But, according to Hempel, this is an instance of an empirical law of coexistence (1965a, 352). How is such a divergence between two philosophical evaluations of the same scientific law plausible? The clue is given in Hempel's study entitled "Theoretician's Dilemma." There he discriminates between two levels of development of a scientific discipline (scientific systematization): the level of empirical generalization (symbolized further on as l_e) and that of theory formation (symbolized further on as l_t) (Hempel 1958, 178). Level l_e is illustrative of earlier stages of a concrete science and is typically concerned with the formulation of empirical laws which relate the immediately observable aspects of an object. In this category Hempel includes, for example, Galileo's, Snell's and Hooke's laws, which, in his view, are simple quantitative laws. Level l_t succeeds level l_e, for at l_t theoretical laws and theoretical principles are formulated which describe the behavior of hypothetical, immediately unobservable entities and processes with the aim of grasping the foundation for those empirical uniformities known from level l_e. I see the scientific law for the simple pendulum as having two forms. It is *first* a scientific law which was formulated *before* the advent of Newtonian mechanics (i.e., before the statement of Newton's laws of motion, of the gravitational law as well as of further derived relations). Given the antecedent idealizations, this law for the simple pendulum could have had, *at most*, the form[3]

$$L^{(6)} : (x)(Px \,\&\, Id_{1,2,4,6\text{-}8}x \rightarrow T^{(6)}x = 2\pi\sqrt{lx/g}) \tag{1.5}$$

Since the pre-Newtonian formula does not embrace in itself any reference to unobservable entities (forces), but expresses only the coexistence of the determinations denoted by T and l, it is virtually an empirical law of coexistence. But as soon as the gravitational law comes to be known as well as the dynamic laws of mechanics, relations for the moment of inertia, moment of force, etc., along with knowledge of non-gravitational forces, then the law for the simple pendulum assumes the form expressed by (1.4) and thus becomes a theoretical causal law. Not only is it formulated as a scientific law with Newton's laws informing it as part of a theoretical background, but it involves the concept of force in a twofold way: in the antecedent idealizations Id_3 and Id_5 and simultaneously in the parameter g in the consequent. At this point g does not come out as an empirically measured constant, as it would in the case of (1.5), but as an acceleration caused by the earth's gravitational field. Its force F is expressible via the equation $F = mg$, where m stands for the mass of the object accelerated by this force. This equation holds rigorously only if two idealizations hold: the accelerated object has a negligible volume, and no forces are acting upon the physical system where this object is located. The second dynamic law of classical mechanics, therefore, appears in the form

$$L^{(2)} : (x)(Ox \,\&\, Id_{1,2}x \rightarrow Fx = mxg^{(2)}x) \qquad (1.6)$$

where O stands for an object with non-zero mass occurring in a certain physical system, and $Id_{1,2}$ stands for the above-mentioned two idealizations. Thus the scientific law for the simple pendulum is a *causal law* which, however, no longer bears the character of a Humean causal law. The point is that it is not a description of the regular succession of observable phenomena but a description of the genesis of oscillations. It is a description of the generative cause, the force F which generates oscillations. This causal law is already a constituent of the level l_t. By contrast, the law expressed by formula (1.5) is still only an empirical law of coexistence and belongs to the level l_e.

A transformation in the construal of causation also happens at the passage from the level l_e towards l_t, where Galileo's law of free fall changes from being a prerequisite of Newtonian mechanics to its consequence. In terms of current physical knowledge, Galileo's law of free fall may be symbolized thus:

$$L^{(9)} : (x)(O'x \,\&\, Id_{1\text{–}9}x \rightarrow s^{(9)}x = g^{(2)}xt^2x/2) \qquad (1.7)$$

where O' means "released body with non-zero gravitational mass occurring in a physical system" and Id_{1-9} stands for the nine further idealizations: 1) the initial velocity of the body equals zero; 2) friction equals zero; 3) the remaining gravitational forces equal zero; 4) the non-gravitational forces equal zero; 5) the earth's volume equals zero; 6) the falling body's volume equals zero; 7) the gravitational force of the earth remains unaltered in time; 8) the earth's acceleration generated by the attraction of the falling body equals zero; 9) the physical system within which the falling body occurs is beyond the influence of any external force (this idealization arises due to the fact that in formula (1.7) there occurs g, which in formula (1.6) is given as $g^{(2)}$, i.e., a transposition of idealizations from the law expressed by formula (1.6) into the law expressed by formula (1.7) takes place). The law expressed by formula (1.7) can now assume the form it might have had prior to the advent of Newtonian mechanics by singling out the idealizations denoted by Id_{3-9}, i.e., it holds[4]

$$L^{(2)} : (x)(O'x \,\&\, Id_{1,2}x \rightarrow s^{(2)}x = gt^2x/2) \qquad (1.8)$$

A scientific law expressed by formula (1.8), as it has been neatly articulated by Hempel (1966, 75), has an empirical character and exemplifies a scientific law of succession, a Humean causal law (Hempel 1965a, 352), since it states that whenever and wherever a body is set free (cause) from the state of rest in the vacuum, it is bound to cover the distance s (effect). The scientific law expressed by formula (1.7), by contrast, is a theoretical law which involves a non-Humean construal of causation. The latter is expressed in it not as a succession of observable phenomena-events of a certain type (set free $\rightarrow$ covered distance), but as a relationship obtaining between the (immediately unobservable) acting force and the result of this action – the covered distance, as a relationship of the generative mechanism to the phenomenon it generated. Though Hempel correctly distinguished level l_e from level l_t still he failed to perceive a crucial change happening at the passage from l_e to l_t, namely the shift in the construal of causation. Scientific laws of the type L_{cs} and L_{cx} as empirical laws are constituents of the level l_e With the development of scientific knowledge to level l_t, they are transformed into theoretical (non-Humean) causal laws. Hempel came to understand (1958, 174; 1964, 291–292; 1965a, 344) that when scientific laws of level l_e are transformed from prerequisites to become consequences of certain scientific laws of level l_t, at that point there occurs a delineation of the range of their validity (i.e., the number of idealizations contained in them increases: in the case of the simple pendulum, from six to nine, and in the case of free fall, from two to nine). What Hempel failed to grasp, however, is that the more rigorous establishment of their validity, or their limits, is rooted in having *introduced into those laws notions which express immediately unobservable entities and processes causing the phenomena under examination.* It is due to this introduction that the laws of the type L_{cs} and L_{cx} are converted into theoretical causal laws, the latter already being constituents of level l_t. Their derivation is accomplished *entirely within level l_t and by no means is in itself a transition from l_t back to l_e.*

1.4 IDEALIZATIONS AND PROVISOS

The reconstruction of the laws for the simple pendulum and free fall has shown that they contain a great number of idealizations. How are these idealizations reconstructed by Hempel? His *first*, and negative, approach to such an application of idealizations was explained in his article (1952). There he adheres to the view that in scientific laws with the logical structure "If *P* then *Q*," *P* cannot represent an ideal type, since the idealizations involved would disable the empirical falsification of such a scientific law. The point is that through incessant *ad hoc* modifications it would be possible to transform a law so that it would always satisfy any empirical test. So, according to Hempel (1952, 168)

> An idealized concept *P* does *not*, therefore, function in hypothesis of the simple form "If *P*, then *Q*" Thus, e.g., the hypothesis "If a substance is an ideal gas then it satisfies Boyle's law," which is of that form, is an analytic statement entailed by the definition of an ideal gas; it cannot serve explanatory purposes.

In his view, Boyle's law should have the structure

$$(x)(IGx \,\&\, Tx = c_1 \rightarrow pxVx = c_2) \tag{1.9}$$

where *IG* denotes the ideal gas; *T* the absolute temperature; *p* the pressure; *V* the volume; and c_1 and c_2 certain constants. In my view, Boyle's law has the structure (*G* stands for gas; V_i for the internal volume of the components of gas; p_i for the pressure between them)

$$(x)(Gx \,\&\, V_ix = 0 \,\&\, p_ix = 0 \,\&\, Tx = c_1 \rightarrow pxVx = c_2) \tag{1.10}$$

and it is a synthetic statement which can serve explanatory purposes. By abolishing the idealizations that posit the gas is ideal, one obtains the law (c_3 stands for a constant)

$$(x)(Gx \,\&\, V_i x \neq 0 \,\&\, p_i x \neq 0 \,\&\, Tx = c_1 \rightarrow (px + p_i x)(Vx - V_i x) = c_3) \qquad (1.11)$$

The fact that idealizations and their abolishment play a fundamental role in scientific explanation can be seen also in the case of the pendulum. The scientific law expressed by formula (1.4), can serve as a point of departure for scientific explanation. In the case of accounting for the pendulum's behavior under the condition of the environment's resistance, it is necessary to suppose that the eighth idealization is no longer true, i.e., to assume that non-Id_8 holds ($\neg Id_8$), contemporaneously presupposing that the volume of the suspended body has a non-zero volume, i.e., that $\neg Id_7$ holds. One can express force F_l counteracting force F as follows. If V denotes the volume of the body suspended and d the density of the elements of the environment per unit of volume, then the mass of the displaced elements of the environment will be $m' = Vd$ and $F_1 = m'g$, while for the period the equation $T = 2\pi\sqrt{ml/F - F_1}$ holds. So from $L^{(8)}$, expressed by formula (1.4), one obtains

$$L^{(6)} : (x)(Px \,\&\, Id_{1\text{–}6}x \,\&\, \neg Id_{7,8}x \rightarrow T^{(6)}x = 2\pi\sqrt{mxlx/Fx - F_1 x}) \qquad (1.12)$$

For a better understanding of Hempel's failure to comprehend the function of idealizations in scientific explanation, let me analyze also R. Carnap's approach to them in his (1939) article where he investigates logical and mathematical calculi as well as their applications in empirical science. The central issue here is Carnap's view of the structure of empirical theories, later labeled *the standard (axiomatic) conception of scientific theories*. According to him, each scientific theory is or should be founded on a logical and mathematical calculus to which a special calculus is added. The latter is built, according to Carnap, by the *top-down* method, i.e., by moving from *primitive signs* and *axioms* towards derived signs and *theorems*. This movement has the character of a *deduction in an uninterpreted calculus* which only by a successive *semantic* interpretation of the derived signs and theorems "is anchored at the solid ground of observable facts" (Carnap 1939, 65). By the direct interpretation of the theorems and derived signs and indirect interpretation of the axioms and primitive signs, the calculus turns into a deductive explanation or prediction of a certain observable event (Carnap 1939, 36). At this point it becomes evident that Carnap did not grasp the importance and function of idealizations in scientific explanation. As an example of a movement in an uninterpreted calculus and of its successive semantic interpretation, he suggests an explanation of the increase of the length of an individual body denoted by c given in a real physical measurement. But as a first premise of his whole deduction there appears the sentence "c is *Sol*" where *Sol* is semantically interpreted as designating solid bodies. But this is equivalent to the patently false claim that in real measurement one deals with solid bodies.

Furthermore, the standard (axiomatic) conception of scientific theories entails the following question. Do the general, abstract terms, obtained by an indirect semantic interpretation of the primitive signs of the special calculus, have an ontological status? Do they denote some really existing entities? Carnap already gave his negative answer in his (1939) article, where he claimed that the semantic interpretation of the primitive signs of the special calculus (e.g., "E" as "intensity of the electric field") is only a movement "from a symbol in a symbolic calculus to a corresponding word expression in a calculus of words" (1939, 68). This view was then expressed explicitly in his later article where he claimed that an "alleged statement of reality of the system of entities is a pseudo-statement without cognitive content" (1950, 214).

Hempel, it seems to me, tried in the fifties to escape the realistic as well as antirealistic answer to the above-mentioned question when he declared that "no generally binding standards can be specified: the answer is ultimately determined by one's philosophical conscience" (Hempel 1958, 210). His evasion of ontological questions in the philosophy of science came to an end in the late sixties. In his (1970) paper he makes an ontic claim about the relation of a theory from level l_t to empirical phenomena expressed in level l_e. "A theory," he argues, "aims at providing a deeper understanding by construing those phenomena as manifestation of certain underlying processes" (1970, 142). This represents a shift away from his earlier epistemic approach, where he claimed that "[s]cientific systematization is ultimately aimed at establishing explanatory and predictive order among the bewildering complex 'data' of our experience" (1958, 177).

This shift is even more visible in his 1988 article, where for the second time he tries to explain the place of idealizations in scientific systematization. Idealizations, labeled by him "provisos," are presupposed in thought movements inside a scientific theory, *C*. If, for example, one presupposes within the electromagnetic theory the proposition "*b* is a magnet," Sc_1, then it should be possible to infer by theoretical principles from *C* that "If *b* is broken into two bars, b_1 and b_2, then both are magnets and their poles will attract or repel each other," Sc_2, (Hempel 1988, 23). But according to Hempel, this inference holds only if the breaking takes place in a certain range of temperatures, since a high temperature would demagnetize both pieces of magnet. Similarly, certain provisos are presupposed in the transition from Sc_2 to a proposition expressed in a vocabulary, *Va*, given antecedent to *C* (i.e., before *C*), e.g., "If *b* is broken into two shorter bars, and they are suspended by long thin threads, close to each other and at the same distance from the ground, then they will orient themselves so as to fall into a straight line," *Sa*, (Hempel 1988, 20). For such a situation to develop it is necessary to presuppose that there is no influence at work, for instance from a strong magnet or an air flow, which would prevent the orientation of the magnets into a straight line. *But*, Hempel claims, *even if provisos have such an important function, they are not and in principle cannot be stated in a scientific theory* for the following two reasons. *Firstly*, their number is potentially infinite. *Secondly*, the provisos very often cannot be stated by means of the theory in question. So, in the case of the law for the simple pendulum in classical mechanics, it is supposed that non-gravitational forces do not act. But, classical mechanics states nothing about this type of force.

The conclusions Hempel arrives at, and which contradict the standard (axiomatic) conception of scientific theories, as well as his own views from the forties to the sixties, are as follows: a) the movement from Sc_1 to Sc_2 and from the latter to Sa does not have a deductive character; b) a refutation of an empirical prediction or a contradiction between an explanation of an event and an empirically established determination of this event, does not as yet refute the theory through which the explanation (or prediction) was made because the refutation or contradiction may have failed to take into account some important provisos; c) due to the possibly infinite types of provisos, each scientific law is vacuously (trivially) fulfilled by each object from the universe for which it is formulated.

In my view, Hempel's 1988 paper again fails to grasp the function and importance of idealizations in scientific explanation. The example of explanation based on the law for the simple pendulum makes it clear that provisos are explicitly formulated inside a theory and very often by means of terms taken from other theories. But, as I shall argue in Chapter 3, the explanation of a universal law by means of another universal law, where both of them contain provisos, does not have the character of a deductive argument. Here I will also try to solve the problem of a non-vacuous, non-trivial fulfillment of a scientific law containing provisos.

But first I will scrutinize the foundations of the D-N model and of the scientific law of appearance. These were explicated in the works of Karl Popper, to which I now turn.

CHAPTER 2

POPPER'S MOA

The aim of this chapter is to reconstruct the concept of natural necessity upon which the empirical causal (Humean) type of scientific law rests. In my opinion, this effort will enable one to grasp the ontological dimension of the epistemological concept: "empirical causal (Humean) type of scientific law." The "empirical causal type of scientific law" was exemplified in the previous chapter by the following two propositions: *Whenever the temperature of a gas increases while its pressure remains constant, its volume increases*; and *Whenever a wire loop is moved across a magnetic field, then current flows in it.*

In this chapter, a new term will be introduced: the term "natural law." One has to bear in mind that the term "natural law" is usually understood as a universal statement or generalization fulfilling certain additional requirements (Popper 1968, 60; Kneale 1961, 63), and in this book that term, defined according to common usage, would be equivalent to the concept of "scientific law." Where necessary, therefore, I will use the phrase "law of nature" or "law of society" to distinguish the ontological concept from its epistemological counterpart, "scientific law."

The starting point of Hume's approach to causation is the supposition that meaningful ideas are analyzable on the basis of the sensory impressions in which they originate, i.e., the content of a concept is given by those aspects of empirical experience in which it originates. In the case of the idea of causation, this supposition leads him to posit three empirical aspects of causation – contiguity, succession (temporal ordering), and regularity (constant conjunction) – as well as its fourth, non-empirical aspect: necessary connection of cause and effect. According to Hume, nothing in reality corresponds to this final aspect of the idea of causation, for it has its origin in our mind and comes into being on the basis of our perception of the regularities.

Hume's view, according to which it is impossible to analyze the problem of causation in terms of objective necessity, found a lot of advocates and was labeled "regularity theory." According to this theory there exists nothing in reality which connects cause and effect, understood as antecedent and consequent events. The regularity theory surrenders Hume's psychological substantiation of the idea of necessary connection. Instead, it grounds this necessity in the logical connection of an individual causal proposition to a general causal proposition, where each causal proposition about a certain individual state of things is logically connected with a general causal proposition. According to regularity theory, what counts in the investigation of causation is not objective necessity, but the universality of causal propositions, i.e., of causal laws. So in this theory priority is given to the explication of the concept "scientific law."

2.1 K. R. POPPER 1934, 1948, 1949

Such an explication in the framework of regularity theory was provided by K. R. Popper in the first edition of his *Logik der Forschung* in the year 1934 and in

(1948). Popper first circumscribes his understanding of causal explanation. According to him (1968, 59–60)

to give a *causal explanation* of an event means to deduce a statement which describes it, using as premises of the deduction one or more *universal laws*, together with certain singular statements, the *initial conditions* . . . We have thus two different kinds of statements, both of which are necessary ingredients of a complete causal explanation. They are (1) *universal statements*, i.e., hypotheses of the character of natural laws, and (2) *singular statements*, which apply to the specific event in question and which I shall call "initial conditions." It is from universal statements in conjunction with initial conditions that we *deduce* the singular statement . . . We call this statement a specific or singular prediction . . . The initial conditions describe what is usually called the "cause" of the event in question. And the predictions describe what is usually called the "effect"'.

The same circumscription is provided by Popper in (1948) where *E* is a singular event-effect to be explained, *I* is the singular event-cause, *i* the statement about *I*, *e* the statement about *E*, and *u* is a scientific universal law. He emphasizes that such an understanding of causal explanation enables one to acquire a new view of "Hume's problem whether there is a necessary connection between a cause and the effect it produces" (Popper 1948, 148). Popper endorses Hume's negative answer "because whenever we consider *I* to be the cause of *E*, we do so in view of a . . . law *u* in the presence of which *e* follows from *i*, and since we may take it that the relationship of deducibility may be described as 'necessary' one, we may say that the connection of *I* and *E* is necessary one . . . necessary relative to *u*" (1948, 148). Popper then in (1968) defines the concept "scientific law" by distinguishing between strictly universal propositions and numerically universal propositions. As examples Popper offers the following statements: "(a) [o]f all harmonic oscillators it is true that their energy never falls below a certain amount (viz. $h\nu/2$); and (b) of all human beings now living on the earth it is true that their height never exceeds a certain amount (say 8 ft)" (1968, 62). According to Popper, even if both statements have the form of a universal material implication, there exists a principal distinction between them. Proposition (Popper 1968, 62–63)

(a) claims to be true for any place and any time. Statement (b) refers only to a finite class of specific elements within a finite individual (or particular) spatio-temporal region. Statements of this latter kind can, in principle, be replaced by a conjunction of singular statements; for given sufficient time, one can enumerate all the elements of the (finite) class concerned. This is why we speak in such cases of "numerical universality". By contrast, statement (a), about oscillators, cannot be replaced by a conjunction of a finite number of singular statements about a definite spatio-temporal region . . . we regard a statement of type (a) as an all-statement.

According to Popper, if one uses the term "scientific law," he has in mind an always true statement (all-statement) because it says "'Of all points in space and time (or in all regions of space and time) it is true that . . .'" (1968, 63) and no proper names occur in it. Popper's regularistic approach to the concept "scientific law" implies its following circumscription. It is a proposition, which 1) has the form of universal material implication; 2) is true in all spaces and times; 3) is as a matter of fact universal, i.e., it expresses a uniform, constant succession of its instances; 4) it is accidentally true, i.e., its truth depends on the state of facts in the actual world (given to us empirically).

2.2 W. KNEALE 1948, 1949, 1950, 1955

W. Kneale in his works (1948, 1949, 1950) stood up against Popper's regularity approach to scientific laws. He asserts (Kneale 1948, 163)

> I do not think that Dr. Popper's account of the matter does justice to our ordinary usage of the phrase "necessary connection." When we say that *E* is necessary connected with *I*, we do not mean merely that *e* follows logically from *i* and some universal statement *u*. We wish to convey that *u* states a necessary connection between kinds of events, and the fundamental problem is to explain the usage of "necessary connection" in this latter context.

He does not endorse the view that scientific laws have the character of a universal material implication, that they express an accidental and constant conjunction of events, and that they are connected to the realm of actual experience. On the contrary, he claims that "our statements of natural law purport . . . to be more than universal material implications, I suggest we should boldly say that they are statements of necessary connections," (Kneale 1948, 164) and "it is therefore a plausible suggestion that they may be *principles of necessitation*" (Kneale 1949, 71). He substantiates his necessitarian approach by the following consideration. Popper's distinction between necessary (lawful) and accidental is based on the distinction of what is valid universally (i.e., at any time and in any space) from what is valid only locally (i.e., in a certain time and in a certain space). Kneale considers the proposition "All ravens are black." From the view of Popper's understanding of the concept "scientific law," there exist only two possibilities: (a) it is a law of nature that all ravens are black, or (b) there exists or will exist in a certain moment of time a raven which is not or will not be black. The accidental is always what is given only locally, while a law is given globally, i.e., "philosophers who treat suggestions of law as universal material implications say in effect that there is no sense in talking of historical accidents on cosmic scale" (Kneale 1950, 48). In his view, if something is universally valid, it does not mean that it is always lawful. "We can conceive an unrestricted material implication without regarding it as a statement of a law," (1950, 49) e.g., "[i]t is at least conceivable that there has never been a chain reaction of plutonium within a strong shell containing heavy hydrogen, and it is also conceivable that there never will be" (1950, 49). One can reason about events which never and nowhere occurred in the whole universe, and still their universal non-occurrence has an accidental character.

W. Kneale unifies his critique of Popper's regularity approach with an explanation of his own necessitarian view. Its basic aim is to strengthen the modality of the concept of scientific law. When one asserts that "All α things are β" expresses a law of nature, then actually he means "All α things must be β" or "α-ness necessitates β-ness" (1949, 78). According to Kneale it says that "it is impossible for an α-thing not to be β" and therefore the idea of natural necessity is at the same time "the notion of a boundary of possibility" (1949, 78). He is anxious to trace the origin of this necessity in the internal characteristic of the objects itself, which are scrutinized by special sciences. In this context he refers to Locke, according to whom "we cannot have insight into the connections which we assert in natural science; but he holds nevertheless that they may be necessary in the same sense as connections which we are able to comprehend, and he implies that a sufficiently strong mind which was furnished with appropriate ideas of what he calls the real internal essence of things might see the necessity of natural laws" (1949, 71).

W. Kneale also quotes scientific theories which refer to unobservable entities (atomic theory, wave theory). Their special feature is that they explain "laws by means of postulates which are not themselves established by direct induction from experience and cannot . . . be tested directly in any way. These postulates are hypotheses about the existence of objects which must, from the nature of the case, be imperceptible . . . the physical world as described in such theories cannot, from the nature of the case, be observed . . . they are concerned with things which are not observed even in principle" (Kneale 1949, 93). Special sciences strive by means of scientific laws "to specify the hypothetical entities in such a way that any connections between them . . . are intrinsically necessary" (1949, 79). At the same time he responds to the question of how one recognizes necessity understood as boundary of possibility, "only by the statement that we know the incompatibility of certain items such as redness and greenness which we have met in experience" (1949, 78).

The result of Kneale's exposition of the necessitarian approach is, in my view, two-fold. On the one hand, he did not succeed in explicating the type of necessity which one deals with in scientific laws and which should reside in the essence of things. He tries to specify this necessity more closely by introducing the modal term "must" and "to necessitate." But the fact is that in Kneale's work these terms are as vague as the terms they are intended to explicate. So it become impossible clearly to identify the distinctive features of this necessity. He states "the notion of necessity used in natural science is the same as used in logic and philosophy, namely the notion of situation without alternative" (1955, 61). Such a situation would be given in each possible world and thus would be a logical necessity. But in a contrary assertion, he states that scientific laws belong to a group of logically accidental statements (Kneale 1962, 649); they do not hold in all possible worlds but only in all worlds of a certain kind (Kneale 1961, 63).

On the other hand, Kneale's deliberations represent an attempt to establish a realistic, non-empiricist foundation for the concept of necessity which I will utilize in the following chapters. In my view, this must be accomplished by using one of Kneale's seemingly marginal ideas about cause and effect. While comparing Hume's understanding of cause and effect with Newton's understanding (especially with his notes on attraction in his *Optics*), Kneale arrives at the conclusion that physics has an understanding of cause different from that of Hume, i.e. *not as an antecedent event* (1949, 60). But, instead of developing such a non-Humean understanding of cause, Kneale satisfies himself by stating that in physics discourse on "cause" one finds no new understanding, but merely a non-philosophical usage of this term; thus, Kneale ultimately accepts the conditional event-analysis of causation and necessity inspired by Hume.

2.3 K. R. POPPER 1958

Popper accepted Kneale's criticism from (1949; 1950) and stated "the characterization of laws of nature as strictly universal statements is logically insufficient and intuitively inadequate" (1968, 428) and that it is necessary to find another criterion to distinguish accidents on a cosmic scale from laws of nature. Popper expounds his revised view in the famous Chapter X* of the new appendices of *Logic of Scientific Discovery* by utilizing the following example. He introduces the universal name, "moa," denoting an extinct species of huge birds which once lived in New Zealand. Here "we ought to admit that it is quite possible – and even credible – that no moas have existed in the universe or will ever exist, apart from those which once lived in New Zealand, and will assume that this credible view is correct" (Popper 1968, 427). Popper asks that we suppose that organisms of the moa type have a certain biological structure which would permit them to live sixty years or even longer, given certain favorable, ideal conditions (e.g., in an environment sterile enough).[1] Popper then suggests the specific conditions under which those birds really lived in New Zealand and which could be far from ideal (e.g., a certain virus was operative there), so that no moa lived longer than fifty years. One obtains the statement

All moas die before the age of fifty years (2.1)

which is true because nowhere and never in the universe has there existed a moa that was older than fifty years. But according to Popper "this universal statement will not be a law of nature; for according to our assumption, it would be possible for a moa to live longer, and it is only due to the *accidental* or *contingent* conditions – such as the copresence of a certain virus – that in fact no moa did live longer" (1968, 427–428). From this example it should be evident, according to Popper, "that there may be *true, strictly universal statements* which have an accidental character rather than the character of true universal laws of nature" (1968, 428). In this way Popper interprets Kneale's terms: "boundary of possibility" and "principle of necessitation." Under favorable conditions it would be possible for individual moas to live longer than the actual moas. So, if there would be a law which would restrict their age to fifty years, "it would not be possible for any moas to live longer. Thus natural laws set certain limits to what is possible" (1968, 428). Popper also implicitly makes reflections about the equation $F = kmm'r^2$ of the law of gravitation which is fulfilled for different individual masses, for different individual distances, and these are termed by Popper "initial conditions." He suggests that we vary "the initial conditions of our world, for example, the distance of planets, their masses, and the mass of the sun," and he then "interprets the result of such changes as the construction of a kind of 'model' of our world (a model or 'copy' which does not need to be faithful with respect to the initial conditions)" and he calls "those statements 'necessary' which are true in (the universe of) *all* these models" (1968, 435–436). In direct connection with such an understanding of the conditions of a scientific law, he provides the following definition of the concept "natural necessity" (1968, 433)

(N^0) A statement may be said to be naturally or physically necessary if, and only if, it is deducible from a statement function which is satisfied in all worlds that differ from our world, if at all, only with respect to initial conditions.

How does the understanding of the concept "scientific law" as expressed here differ from his 1934 understanding of this concept? First of all, the definition has undergone a substantial shift, moving away from the requirement of universality, space-time unlimitedness, towards the requirement of complete unconditionality, or at most, as it is stated in *(N^0)*, the dependence on singular conditions, i.e., the independence from other types of conditions as singular conditions.

Popper's 1959 view implicitly contains a new understanding of the structure of scientific laws. Namely, that a scientific law always contains the delimitation (specification) of the type of structure for which it is formulated,[2] and it is completely irrelevant whether the instances of this structure are given in the whole universe or just in some part of it.

A scientific law does not make a statement about all regions of space and time, but only about all those regions where instances of that structure exist. So, for example, the equation $E = h\nu/2$, which Popper mentioned in his 1934 approach, does not make a claim about all regions of space and time, but only about all those regions where oscillators of a certain type exist. Similarly, the proposition about the maximum height of human beings now living on earth cannot be denied the status of a scientific law just because human beings exist only in a certain region of space and time.

Simultaneously, Popper refuses to accept Kneale's necessitarian approach as well as his attempts to link logical and natural necessity in some way. According to Popper "compared with logical tautologies, laws of nature have a contingent, accidental nature. For there may be structurally different worlds – worlds with different laws" (1968, 430). This in fact means that Popper identifies logically possible worlds with possible worlds. Only through this identification, together with the reasonable supposition that in structurally different possible worlds different laws of nature are at work, can one arrive at the above-stated comparison between logical tautologies and laws of nature.

But the whole controversy between the regularists and the necessitarists starts from this identification. The latter is based on the Humean supposition that logical necessity is the only existing type of necessity, whose origin is based in our way of defining terms appearing in logical tautologies. Popper confirms this Humean standpoint by stating "I regard, unlike Kneale, 'necessary' as a mere word – as a label useful to distinguish *the universality of laws* from 'accidental' universality. Of course, any other label would do just as well, for there is not much connection here with logical necessity. I largely agree with the spirit of Wittgenstein's paraphrase of Hume: 'A necessity for one thing to happen because another has happened does not exist. There is only logical necessity'[3]" (Popper 1968, 438). If one introduces the symbol "N" as a name of the class of propositions which are physically necessary in the sense of the definition *(N^0)*, then it is possible to define "a -N->b" ("If a then necessarily b") as follows[4]

$$a \text{ -}_{\text{N}}\text{-> } b \text{ is true if and only if } (a \rightarrow b) \in \text{N}$$

According to Popper "[o]nly in one way is a -N->b connected with logical necessity: the necessary link is neither to be found in a nor in b, but in the fact that the corresponding ordinary conditional . . . follows with *logical necessity* from laws of nature – that it is necessary, relative to a law of nature" (1968, 438). In this argument he also finds confirmation of his views expressed in (1948). "[I]t may be said that a law of nature is necessary in its turn because it is logically derivable from, or explicable by, a law of a still higher degree of universality . . . One might suppose that it is this logically necessary dependence upon true statements of higher universality, conjectured to exist, which suggested in the first instance the idea of 'necessary connection' between cause and effect" (1968, 438–439).

2.4 MOA AND ITS CRITICS

After the publication of the definition *(N^0)* by Popper in 1959, several papers appeared which dealt critically with his new approach to scientific laws and natural necessity. The most important of these, in my view, is the paper by Nerlich and Suchting (1967). The authors introduce the name *M* for the statement (2.1) under the same supposition as it was introduced by Popper. They also introduce the statement *C* about certain conditions (e.g. presence of a virus) and a statement *L* about the structure and functioning of organisms of the moa-type. On the one hand, *M* is not a scientific law because it is not invariant to changes of conditions *C* (presence/absence of the virus). On the other hand, *L* is a scientific law,[5] because its truth is independent of changes of C. Therefore, *(N^0)* can be expressed as follows: "A statement may be said to be naturally or physically necessary if, and only if, it is deducible from a statement function which is satisfied in all worlds which instantiate the properties and relations of our world but which may differ from our world with respect to initial conditions" (Nerlich and Suchting 1967, 64). But the following question then arises. What is the admissible range of the variations of initial conditions? The authors prove (1967, 64) that if Popper's view should be different from Kneale's, then the only admissible class of initial conditions is that of physically possible initial conditions. For the elements of this class, it holds true that they are not inconsistent with the physically necessary proposition, i.e., with the scientific law. But then it becomes apparent that Popper's definition *(N^0)* is circular because it states that a statement may be said to be naturally or physically necessary if, and only if, it is deducible from a proposition function which is satisfied in all worlds which have the same laws as our world.

If *(N^0)* is circular, then consequently Popper's understanding of the concept "accident" is also circular. He understands an accident as something given by the intersection of a law of nature with conditions which, with respect to it, do not have the character of initial conditions. But just as it is impossible for him to define "law of nature" in a non-circular way, so too it is impossible for him to determine the meaning of "conditions which with respect to a law of nature have the character different from its initial conditions."

The problem of finding a non-circular definition of "scientific law" is not a problem confronted exclusively by Popper. It was also encountered by W. Kneale, who understands this term as "a generalization which holds for all possible worlds of some kind" (1961, 63). Here again the question arises: is it possible to select this kind of world – i.e., to differentiate worlds of this type from worlds of a different type – without using the phrase "worlds which have the same laws of nature as our world?" I shall outline some initial intuitions at the end of this chapter.[6]

2.5 G. N. SCHLESINGER AND SORTS OF INITIAL CONDITIONS

An interesting attempt at a definition of the concept of scientific law starting from Popper's definition *(N^0)* is provided by G. N. Schlesinger in his (1987). Let me take the statement S

All the fruit in Smith's garden are apples (2.2)

Even if this generalization states certain facts given in a limited space-time region, still it is possible that it expresses a really existing law. One can imagine the following scientific law S_1

All the fruit in any garden with property P are apples, except that certain factors prevent the development of more than one such peculiar system. (2.3)

and the fact that nowhere, except in Smith's garden, has this state been obtained is only a consequence of the fact that in all other gardens a certain external interference (e.g., a destructive insect) prohibited the growth of apples. But it is equally possible that the fact stated in S is not a law but only a consequence of an accidental intersection of conditions just in the place where the garden of Mr. Smith is located. According to Schlesinger, the definition *(N^0)* creates, even if it is circular, the prerequisite for a distinction of the two cases when S is true in the actual world. For such a distinction he first introduces two supporting statements. "(1) If strict determinism is true, then a full description of the initial conditions implies the complete set of laws that govern nature. (2) Two initial conditions in two universes may be non-identical yet of the same *sort*. Two initial conditions prevailing in two universes are of the same sort if all entities they contain – which may differ in number and spatial configuration – are of the same sort," (1987, 383–384) and where "entities a and b are of the same kind (or sort) if the only difference between them is their spatial and temporal position" (1987, 384). Then it is possible to formulate the conditions of necessity for S. Let me imagine a situation where S is true not only in the actual world but also in all other worlds which have the same kinds (sorts) of initial conditions (even if they differ by their respective initial conditions). All these worlds behave according to the same laws as our actual world; S is here completely determined by laws of nature and is therefore physically (naturally) necessary.

Let me imagine another situation when S is true only in the actual world, but not in all other worlds which have the same kinds (sorts) of initial conditions, i.e., the same laws of nature as our world. The falsity of S in these worlds is only a result of the existence of certain values of the initial conditions. In both cases, everywhere where S is true, it will also be physically necessary. On the basis of these considerations Schlesinger provides the following definition (1987, 385):

(PN) A statement is physically necessary if and only if it is deducible from a statement function which is satisfied in all worlds that may differ from our world but which have the same sort of initial conditions.

In his view *(PN)*, unlike *(N^0)*, is no longer circular because the set of worlds for which natural necessity is defined is understood as "having members that do not differ at all in kind of entities they contain at some time; they may differ, however, in the quantity of these particulars and the way they are distributed and arranged in space" (1987, 385).

On the basis of *(PN)* Schlesinger distinguishes two cases of the truth of *S*. (1) The truth of *S* is not generated by laws of nature alone, regulating the behavior of entities with property *P*. Under different external conditions (absence of certain climatic conditions, different behavior of Smith's family) *S* would not be true. So "*S* provides no description of any of the intrinsic features of *P*, it is not inherent in the character of Smith's garden to produce no other fruit than apples. What did constrain the garden to produce apples and nothing else, were external conditions that are not implied by the statement describing *P*" (1987, 386). In this case *S* does not express a law of nature but only an accidental state of things. (2) There exists a direct causal connection between the law of nature for *P* and the fact that all fruit in Smith's garden are apples. Irrespective of the external (climatic, social, etc.) conditions, all the fruit in Smith's garden will be apples. In this case *S* expresses a law of nature.

Does Schlesinger's definition *(PN)* overcome the problems of Popper's definition (N^0)? Let me first examine cases (1) and (2) for Smith's garden from the view of natural necessity. The difference between the cases when *S* expresses (does not express) a law of nature is based on the idea that the state of things expressed in *S* is generated only by (or also by something else than) a law of nature. The definition can be stated as follows:

A certain proposition about a state of things expresses a law of nature if and only if this state is generated by the law of nature alone.

In my view, the definition *(PN)* itself is also circular. What does the phrase "same sort of initial conditions as our world" really mean? If I take as an example the equation from the law of gravitation, implicitly used by Popper (1968, 433) $F = kmm'/r^2$, then I have here the type of condition "non-zero gravitational mass of a body" as well as "the distance between two bodies with non-zero gravitational masses." Regarding their relationship to the law of gravitation, the possible worlds will differ from each other by their respective singular conditions, but these two types (sorts) of initial conditions will be common to all of them. These worlds will be inhabited by entities with a non-zero gravitational mass which exist respectively in a mutual (non-zero) distance. But when we speak about "those possible worlds which have the same sorts of initial conditions as our actual world," we do that always with respect to a statement which, we hope at least, expresses a law of nature. So, for example, in the case of the law of gravitation I construct possible worlds as copies of our actual world by populating them with entities whose properties are known directly from the statement about the actual world. So, in my view, if one uses the phrase "worlds which have the same sort of initial conditions as our actual world," then he is only restating the phrase "worlds which have the same laws of nature as our world."

The conclusion I arrive at, then, is that Popper's definition (N^0) and Schlesinger's definition *(PN)* are equal, and both are circular. This circularity also manifests itself when one confronts *(PN)* with the following Nerlich-Suchting-style question: What are admissible sorts (kinds) of initial conditions? Schlesinger provides the answer to it in a concealed form in his final definition: "A statement is . . . physically necessary, just in case it is not true in all possible worlds, but [is true] in all worlds governed by the same laws as our world, even though those worlds may vary as widely as the laws permit them to vary" (1987, 390). So, the only admissible sorts of initial conditions are those which are not inconsistent with scientific laws.

Schlesinger's approach, even with this shortcoming, represents an important attempt at a definition of the concept of "scientific law." To this end, it introduces the concept "sorts (kinds) of initial conditions" by referring to "entities of a certain kind," i.e., to something which Popper labeled in 1959 by the name "moa." I will take the advantage of this approach in subchapter 2.7, where I will scrutinize propositions like "All moas in favorable conditions will, due to their biological structure, live sixty or more years" and "All the fruit in any garden with property *P* are apples, except that certain factors prevent the development of any such peculiar system."

2.6 J. L. MACKIE AND THE PURIFICATION OF A SCIENTIFIC LAW

Kneale's necessitarian approach was taken up by J. L. Mackie in his (1980). The latter rejects Popper's regularity approach to the concept of natural necessity by referring to distinctions between what is valid globally and what is valid only locally. With respect to Popper's "moa" example, he maintains that it is "not yet clear . . . why does the biological structure of the moa contribute to laws, but the presence of virus only to an accidental universal" (1980, 209). In his view the statement (2.1) can also be understood as a scientific law when we introduce into it a clause stating the presence of a virus. For a better comprehension of such an adaptation Mackie introduces the terms "laws of working" and "collocation"[7] in the following example. Let us have on the surface of the earth an inclined plane, by angle *k*. On its upper edge three red balls (R_1, R_2, R_3) and three blue balls (B_1, B_2, B_3) are placed, a few inches apart from left to right in the order $R_1B_1R_2B_2R_3B_3$. After their release, they will run down the inclined plane and pass a certain imagined point *P*. Let us scrutinize the following proposition (Mackie 1980, 210):

Wherever there is a red ball there is a blue ball immediately to its right. (2.4)

Whenever a blue ball passes P a red ball passes shortly afterwards. (2.5)

Whenever a ball is released on this plane it runs down in accordance with the law $a = \mathbf{g sin} k$. (2.6)

Whenever an object of mass m is near the surface of the earth it is subject to a force **mg** *in the direction of the center of the earth.* (2.7)

*Whenever an object of mass m is subject to force f it is has an acceleration **f/m** in the same direction as that of force.* (2.8)

Between any two objects there is a force proportional to the mass of each and inversely proportional to the square of distance between them. (2.9)

According to Mackie (2.4) represents a true physically accidental proposition, where its accidentality is not a consequence of its space-time confinement. (2.5) is a derived law obtained from (2.6) in conjunction with the collocational proposition (2.4) and the description of the initial set-up of the experiment. But (2.6) is also a derived law obtained from (2.7) and (2.8). Only (2.8) and (2.9) express basic laws of working. Still, Mackie is convinced that it is possible to purify (2.5), (2.6), and (2.7) by substituting in them those parts which express singular conditions of the experiment by general terms, and so change them into pure, though derived, laws of working. So, e.g., in (2.7), the term "earth" is replaced by "a sphere of such a radius, mass and so on." It means "we purify the mixed laws of their collocational component by putting it explicitly into the antecedents of the conditionals" (Mackie 1980, 211). Similarly, the proposition "All moas die before the age of fifty," expressing the collocation of a virus and of moas in New Zealand, is in fact a mixed product which can be also purified by introducing the clause "in the presence of a virus of such and such characteristics." Then, according to Mackie, Popper's definition *(N^0)* can be restated as follows (1980, 212):

[C]alling naturally or physically necessary all pure laws of working, basic or derived, and all singular conditional statements which these entail.

But when one introduces the concept "law of working" it is necessary, according to Mackie, to answer the following question. "Why should laws of working be said to have natural *necessity*? . . . [I]s there anything more about laws of working that would justify their being singled out as necessary?" (1980, 212, 214). Kneale's approach, making reference to logical necessity, to necessary relations of unobservable entities, and to incompatibility of colors, is characterized by Mackie as rationalism in the tradition of Locke and Descartes (1980, 12, 214, 223). On the one hand, he accepts Kneale's idea that in cognition one has to "advance beyond a view of causation as mere regular succession when we conjecture that there really is some causal mechanism underlying the succession and explaining it" (1980, 216–217). On the other hand, he refuses to use transcendent hypotheses and states "would it not be even better actually to uncover and observe the mechanism?" (1980, 217). Mackie claims that "cause" as defined by Kneale is something unobservable which has a certain structure which changes into some such other, also unobservable, with a similar structure, and this is already the effect. One can imagine a certain wave structure changing into another wave structure, where it is possible to find for both of them certain identical values for some of their determinants (number of particles, total energy, etc.). So, if one speaks about necessity on the level of basic laws of working, then he really means "forms of persistence . . . the persistence of objects and the persistence of self-maintaining processes, saying that in both of these we could regard an earlier phase as a cause and a later phase as an effect" (Mackie 1980, 221). Such a necessity provides the ground of causal processes at the observable level, where after the cause a completely different effect follows. So, for example (Mackie 1980, 221–222):

[when] a match is struck on a matchbox and a flame appears: on the face of it this effect has nothing in common with its cause. But if we were to replace the macroscopic picture with a detailed description of the molecular and atomic movements with which the perceived processes are identified by an adequate physico-chemical theory, we should find far more continuity and persistence.

Instead of a rationalist identification of necessity with logical necessity and incompatibility of colors, Kneale's transcendental hypotheses, according to Mackie, "reveal only continuity and persistence," (1980, 223) which he understands as an "empirical counterpart of rationalists' logical or near-logical necessity" (1980, 228). Mackie, then, arrives at the following definition: "A singular causal sequence instantiates some pure law of working which is itself a part of partial persistence; the individual sequence therefore is . . . identical with some process that has some qualitative or structural continuity," (1980, 229) and where "the pure laws of working . . . are never the subject of observation, but are reached by the framing and testing of hypotheses, by the gradual sorting out of these laws from what depends upon collocations or initial conditions" (Mackie 1980, 230).

2.7 SOLUTIONS AND PROBLEMS

Let me examine the following proposition:

All moas under favorable conditions will live due to their biological structure sixty years or longer. (2.10)

Popper has shown in his analysis that the concept "very favorable conditions" (e.g., with respect to the actual state in the swamps in New Zealand – the absence of a virus dangerous to the moas) does not correspond to the concept of initial conditions. But his successive deliberations, as well as his definition *(N^0)*, contain only the concept of initial conditions. This lack is surprising, but what is even more surprising, Popper does not even explicate in (1968) the concept "initial conditions." Our task, therefore, will be to reconstruct such a type of scientific law, a scientific law which would express the knowledge of conditions different from singular conditions, and to explicate, by comparing various types of conditions of a scientific law, the concept of singular conditions. As an example, let me express proposition (2.10) by the following formula:

$$(x)(Mx \,\&\, Id_{1-l}x \rightarrow A^{(l)}x = f(Ix)) \tag{2.11}$$

M denotes a class of organisms of the moa type; Id_{1-l} denotes a conjunction of 1 to l ideal conditions; $A^{(l)}$ denotes the age of moas under these ideal conditions and acquires values of at least sixty years; $f(I)$ denotes a certain function of the internal (biological) structure of organisms of the type denoted by M. Thus the formula (2.11) expresses Mackie's idea of the purification of a mixed product, but it moves beyond Mackie's proposition in at least two central issues. Firstly, I do not take the collocations as initial conditions of a singular event, but as conditions related to the members of the class considered by the law.[8] Secondly, the knowledge about these conditions is expressed not only as knowledge about their impact, working in the antecedents of the derived laws of working, but also, and primarily, in the antecedents of the basic laws of working, as thought-supposition about the non-existence, non-action of certain conditions.

On the basis of (2.11) it is already possible to understand proposition (2.1) about the moas in New Zealand as a scientific law, i.e., stating a result of the action of a virus of a certain type (for example attacking the brain of the moas), which can be expressed as follows:

$$(x)(Mx \,\&\, Id_{1-k}x \,\&\, \neg Id_l x \rightarrow A^{(k)}x = g(I'x)) \qquad (2.12)$$

Id_{1-k} denotes that 1 to k idealizations still hold; $\neg Id_l$ denotes that the l-th idealization does not hold any more. i.e., that the virus is at work already. $A^{(k)}$ denotes the age of moas living in these conditions and acquiring values of at most fifty years. $g(I')$ denotes a function of the modified, with respect to the original (I), biological structure of the organisms of the moa type, where this modification set in due to the action of the virus of the given type. (2.12) expresses the intersection of the law of nature expressed in (2.11) with the working of that type of a virus. The very act of their intersection might have happened accidentally, but once it happened, then the consequent death of the whole moa population at the age of at most fifty years was lawful and necessary (e.g., the virus was passed from generation to generation). So, in my view (2.12) expresses a scientific law even if its non-vacuous (non-trivial) fulfillment depends on the existence of conditions different from singular conditions.

Formula (2.11) raises the following fundamental question. For what types of objects is the scientific law, expressed by this formula, fulfilled, if in our actual world all moas behave according to the law expressed by (2.12)? To answer this question it is convenient to use the concept of possible worlds. In my view, scientific laws of the type (2.11), i.e., those containing idealizations, are *firstly* a result of the cognition of our actual world. From here, then, the "construction of possible worlds" begins. The whole universe can be divided into the set M, containing all the possible worlds in which entities of the moa type exist, and into the set $\neg M$, containing all the possible worlds in which entities of the moa-type do not exist. If, for the sake of simplicity, I put $k = 1$ in (2.11), then the set M can be divided into two subsets. In one, W, the favorable (ideal) condition for the life of moas is given; in the second, W', the non-favorable condition (the virus) is present. W, and W' respectively, are further divided into singular worlds. In each singular world from W a moa will live at least sixty years, while in each singular world from W' it will live at most fifty years.

Each of these "facts" about the singular world are lawful not only inside one and the same subset, but also with respect to the individual worlds from the other subset, even though these subsets, and therefore also their respective singular possible worlds, differ from each other not only with respect to singular conditions. Both subsets W and W' are lawful; and the origin of this can be traced to the common basis upon which these worlds are built. In worlds constructed on such a basis, the existence of entities of a certain kind M is presupposed, which are characterized by a certain (biological) structure I due to which they have a certain life expectancy. So formula (2.11), in the context of Kneale's and Mackie's ideas about the generative mechanism and the form of persistence, is non-vacuously (non-trivially) fulfilled by all those moas whose age was in thought – by neglecting all secondary conditions – grounded on its basic, principal cause, on the corresponding, for the moas, typical, biological structure.

Scientific laws expressed in (2.11) and (2.12) are, of course, vacuously (trivially) fulfilled in the set $\neg M$. Does this fact contradict the basic intuitions connected with the concept "scientific law"? In my view there is no contradiction because a scientific law is always formulated in such a way as to be directed at objects of a certain kind. So, for example, in the case of an empirical (Humean) causal type of a scientific law, the latter states that if something is of a certain kind (wire loop, gas, etc.) and if it undergoes a change of a certain type (movement in a magnetic field, heating, etc.), then it will undergo another type of change (current will flow in it, expansion will set in, etc.) (see, e.g., Will 1984). Such an approach to laws of nature enables one to examine in a different way Popper's claim that logical and physical necessity are different because there are structurally different worlds with different laws of nature (Popper 1968, 430). From the fact that there exist, in the subset $\neg M$, other structures than moas obeying laws different from that expressed in (2.11) and (2.12), it does not follow that (2.11) and (2.12) are false in this subset. They are true, and their truth is here decided only on the basis of an *a priori* knowledge of semantic rules in the same way as the truth of the statement "All bachelors are unmarried." And it does not follow either that these laws could not express a necessary connection in the subset M, namely a connection between a being of the moa-kind and the biological structure denoted by I. So as (2.11) is vacuously fulfilled in $\neg M$ and W', the only (and most important) worthwhile task left to do for science is to analyze this relation in the world W.

It is exactly in this direction that the article of M. Fisk (1970) aims. He states that "physical necessities involve three things, two properties and a kind. The two properties are necessarily connected as properties of things of that kind, but not as properties of a different kind" (1970, 388). For a necessary connection, symbolized as N_l, of properties F and G it does not hold $(x)N_l(Fx \rightarrow Gx)$ but $(x)N_l(Nx \rightarrow (Fx \rightarrow Gx))$. "It may be the case that F is not necessarily connected with G in all possible worlds simply because the connection fails in worlds in which there are things other than things of kind N with F " (Fisk 1970, 388). These ideas were further developed by D.-H. Ruben in (1979). In his view the necessary connection of F and G with N is given in each possible world "only if F and G are properties which are essential for things of that kind. That is, assume that there is some distinction to be drawn between properties which things of a kind have essentially, and properties which things of a kind have non-essentially. Thus, if F, or G, or being G if F, is a property which a thing of kind N has essentially, then if a thing is of that kind, then it is F, or G, or being G if F, in all possible worlds. But the same result certainly does not follow if these properties are held only non-essentially by things of kind N" (Ruben 1979, 64). A scientific law which does not express a relation between essential properties of entities of a certain kind is not true even "as applied only to things of that kind in all possible worlds" (Ruben 1979, 65).

The moral of the possible-world approach to scientific laws can be summed up as follows. The opposition of logical and physical necessity with respect to scientific laws is based on a wrong understanding of their structure. Since they are always formulated with respect to entities of a certain kind and certain favorable/non-favorable conditions, it can be readily seen that the implicit initial ambition in the formulation of a scientific law is that it should be (non-trivially) *true* in those possible worlds where these entities and conditions are given. Therefore, the formulation of a scientific law represents in fact an attempt at a formulation of a logically necessary statement.

By way of conclusion to Part 1, let me return to its very beginning. The starting point was an attempt to analyze scientific laws by means of the formula (1.2). This formula was exemplified by propositions like "All gases expand when heated under constant pressure" or "All metals expand when heated." In (1.2) C^o and E^o were interpreted as denoting types of events which, in accordance with Hume, qualify as cause and effect with no necessary relation between them. Their connection is here being expressed by a sign of material implication, i.e., this sign communicates a certain understanding of causation. But when one studies formula (1.6) in relation to the second dynamic laws of classical mechanics, one readily sees that the relation of the cause and effect is expressed by the equation in the consequent. Contrary to the pattern that characterized the law of the expansion of gases, cause and effect no longer appear as antecedent and consequent events of a certain type. No expression in the antecedent of (1.6) denotes the cause, and the equation as a whole in the consequent does not denote the effect. Therefore, the sign of implication in (1.6) does not express any causal sense. The conclusion I arrive at here is that in some "part" of the special science, laws are formulated whose structure can be expressed by formula (1.2) and whose understanding of causation corresponds to the Humean understanding. But in the case of the second dynamic law of classical mechanics, formula (1.2) no longer corresponds to it; the understanding of causation expressed in it does not correspond to a Humean interpretation. Is this conclusion surprising at all? To answer this question one has to understand the aims pursued by all those philosophers who tried, and still are trying, to grasp the meaning of "scientific law" through a conditional analysis. Their aim was, and still is, to isolate and to define the term "law of nature," through an analysis of the most successful, contemporary scientific laws. By utilizing the way that being "shines through" in our scientific knowledge, they hope to formulate a "pure" ontology. But when one compares the relevant propositions, e.g., the structure of the proposition for the expansion of gases with formula (1.6) for the relation of force and acceleration, he finds out that they contain a different comprehension of causation; they are carrying in themselves implicitly different ontologies. They are statements about reality; and as statements they are human creations; they state the subject's view of the world. A pure theory of being – ontology – which would not involve (at least implicitly) an epistemic horizon of the subject of cognition is a pure fiction.

This fiction is connected with the following illusion. Usually it is supposed that a scientific law *has to* express all the relations between the determinations of the kind of object for which it is formulated; that it expresses all its relations as necessary ones. But, in my view, because any scientific law is a human creation, it is formulated inside a certain horizon of knowledge, inside a certain conceptual network, and therefore has always only a certain conceptual "outfit" at its disposal; for this reason, it can express at a certain stage of its development very often only certain relations between the determinations of the objects of the given kind *N*. A number of theorists have failed to take this fact into account; for instance, J. L. Mackie ignores it in his discussion of the balls on the inclined plane. In his opinion, if we replace the term "earth" in (2.7) by the term "sphere of such-and-such radius, mass and so on," then we obtain a pure, non-collocational law which is already "a mathematical consequence of the basic laws alone" (Mackie 1980, 211).

The same task could be accomplished, according to Mackie, with (2.6) also if "we insert this relevant description of the earth and a description of the relevant features of the plane" (Mackie 1980, 211). Mackie does set limits by warning that this purification can be accomplished "only where we take it that we fully understand how the system in question works, as we do with Newtonian mechanical systems" (1980, 211). But can scientific laws purified in such a way be pronounced completely free of all contingencies; can they be rendered completely "pure," and are they only mathematical consequences of (2.8) and (2.9)? By purifying (2.7) in the way proposed by Mackie, one would obtain a proposition that still contains an accidental, collocational moment, namely the moment indicated by the phrase "near the surface." From (2.8) and (2.9) it does not follow at all that something is near the surface of something else. Therefore, in such a way, purified law cannot be regarded only as a consequence of (2.8) and (2.9). The same also holds for the purification of (2.6). One would obtain a proposition containing, with respect to (2.8) and (2.9), the collocational moment, taken from the purified (2.7), that the ball is near the surface of earth. It would also contain, with respect to (2.7), two new collocational moments, namely that the ball moves on a *plane* and that this plane is *inclined*.

So, again a purified (2.6) is not solely a mathematical consequence of (2.8), (2.9) and (2.7). But the passage from (2.8) and (2.9) towards a purified (2.7), and therefore also towards a purified (2.6), has a collocational character due to another reason completely neglected by Mackie. To obtain the relation *mg* in (2.7), and subsequently the symbol *g* denoting acceleration in (2.6), it is necessary to presuppose that the gravitational mass from (2.9) *equals* the inertial mass from (2.8). But classical mechanics does not theoretically substantiate this equality. Their equality is not a necessary consequence of the relations inside (2.8), (2.9) or between them. Therefore, the passage from (2.8) and (2.9) towards (2.7) does not bear the character of a pure mathematical derivation – deduction. In fact the opposite is true. The more we move away from the "core" of classical mechanics, represented by (2.8) and (2.9), towards its "peripheries" (law of free fall, simple pendulum, movement on an inclined plane. etc.), the more new collocations intervene, i.e., this movement toward the "peripheries" increasingly vitiates the character of a mathematical derivation. The body of knowledge inherent in the Newtonian mechanical system is essentially incomplete because it does not explain all the relations and properties of objects it scrutinizes as necessary, but still it can be said to represent scientific knowledge, even if only a specific type of it, which need not be the ultimate one.

The conclusion of Part 1 can be stated as follows. The concept of an empirical (causal as well as non-causal) type of scientific law, as well as the D-N model, can be used for a philosophical analysis of certain scientific theories, but they must not be absolutized so that they become the only accepted "tools" for such an analysis. The next task is to formulate the structure of a theoretical law of science based on a non-Humean understanding of causation and thereby to create a new model of (theoretical) explanation which would correspond to real scientific explanations better than the D-N model does. Such a formulation should incorporate Kneale's idea about hypothetical entities and Mackie's suggestions about generative mechanisms, forms of persistence, and the purification of scientific laws. Also it should incorporate the concept of conditions different from the concept of singular conditions, as well as the concept of essence of a natural kind discussed by Fisk and Ruben. The best starting point for the realization of this task appears to me the philosophy of science created by the Poznań school, most notably by the foundational work of Leszek Nowak.

PART 2

THE IDEALIZATIONAL LAW OF ESSENCE

The truth of *being* is *essence*. Being is the immediate. Since knowledge wants to recognize that truth of what being is *in and for itself*, it does not stop at the immediate and its determinations, but penetrates it, on the supposition that beyond this being there is something other than being itself, that this background constitutes the truth of being (Hegel 1923, T. 2, 3; 1969, 389).

. . . from the phenomena of motions to investigate the forces of nature, and then from these forces to demonstrate the other phenomena . . . (Newton 1946, XVIII–XVII; 1687, A2).

CHAPTER 3

THE IDEALIZATIONAL LAW OF ESSENCE

3.1 L. NOWAK AND THE IDEALIZATIONAL TYPE OF SCIENTIFIC LAW

The task of this chapter will be to outline the foundations of the idealizational philosophy of science as it has been presented by one of the main representatives of the Poznań School, Leszek Nowak. I shall proceed by juxtaposing his works (1971; 1972; 1974; 1977; 1980; 1985) with physical theories and Marx's economical works. This will enable me to identify the significant benefits as well as the limitations of this trend in the contemporary philosophy of science.

L. Nowak bases his reflections in (1980) on the explication of the scientific law of value in Marx's *Capital.* He argues that for Marx, "the prices of commodities are formed with respect to their values (the amount of socially indispensable time to produce these commodities). But that holds only under certain conditions," (1980, 3), which are eight in number. Following Nowak, let me put them down in relation to the individual variable x as follows:

$$PCx$$

"The commodity x is produced in the conditions of perfect competition," (1980, 7–8) that is, there is no monopolistic domination; it will be labeled as $p_1 = 0$.

$$Ex - Ix = 0$$

"[T]he difference between the exports and the imports of a given economic system producing the commodity x equals zero," (1980, 8). I will label this as $p_2 = 0$.

$$m'x - M'x = 0$$

"[T]he rate of surplus value characteristic of the capital used to produce the commodity x is identical with the average rate of surplus value characteristic of the capital engaged in the economic system in question as a whole (since the value of the difference of these magnitudes equals zero)," (1980, 8). I will label this as $p_3 = 0$.

$$ORG_A x - ORG_I x = 0$$

"[T]he difference between the organic composition of agricultural capital and the average organic composition of remaining capital in a given economic system producing the commodity x equals zero," (1980, 8). I will label this as $p_4 = 0$.

$$merx = 0$$

"[T]he merchant profit on commodity x equals zero," (1980, 8). I will denote this as $p_5 = 0$.

$$ORG_B x - ORG_E x = 0$$

"[T]he difference between the organic composition of capital in that branch of production which turns out the commodity x and the average organic composition of capital in the whole economy in question equals zero," (1980, 8). I will label this $p_6 = 0$.

$$Dx - Sx = 0$$

"[T]he difference between the supply of the commodity x and the demand for that commodity equals zero," (1980, 8) a relationship which I will designate as $p_7 = 0$.

$$Orgx - ORGx = 0$$

"[T]he difference between the organic composition of the capital used in the production of commodity x and the average organic composition of capital in that branch of production which turns out the commodity x is equal zero" (1980, 8). I will label this as $p_8 = 0$. Simultaneously, L. Nowak introduces the condition p_0, which is of a completely different type compared with the above-mentioned conditions, but which has been implicitly embraced within them. The object x is always a commodity; it is what can be expressed as Cx. Now, according to Nowak, the scientific law of value, involving eight idealizations, can be expressed as follows:

$$L^{(8)} : (x)(Cx \,\&\, p_{1\text{–}8}x = 0 \rightarrow P^{(8)}x = l(Vx)) \qquad (3.1)$$

where $p_{1-8} = 0$ means the conjunction of the idealizations mentioned above; $P^{(8)}$ the price of the commodity when all these idealizations hold; V stands for value and l denotes a linear function.

Nowak proceeds to argue that Marx keeps the quoted formulation of the law of value intact until Chapter X in Volume III of *Capital* (i.e. until the third "subchapter" of the second chapter in the manuscript of this volume), where he removes the idealization $p_8 = 0$, i.e., he replaces it with the condition $Org - ORGx \neq 0$. This enables him to introduce a distinction between the individual value and the market value of a commodity. Removing $p_8 = 0$ changes $L^{(8)}$ into $L^{(7)}$ ($p_{1-7} = 0$ means the conjunction of seven above given idealizing assumptions)

$$L^{(7)} : (x)(Cx \,\&\, p_{1\text{–}7}x = 0 \,\&\, p_8x \neq 0 \rightarrow P^{(7)}x = l(Vx) + a(Orgx - ORGx))$$

where the function $a(Orgx - ORGx)$ satisfies the ensuing three conditions:

1. $a = 0$ if $Orgx = ORGx$
2. $a < 0$ if $Orgx > ORGx$
3. $a > 0$ if $Orgx < ORGx$

and $L^{(7)}$ is the scientific law of market-value. Subsequently Nowak points out that by removing the idealization $p_7 = 0$, that is, if $Dx - Sx \neq 0$ holds, it is possible to account for the deviation of the market-price from the market-value. $L^{(7)}$ is transformed into:

$$L^{(6)} : (x)(Cx \,\&\, p_{1\text{–}6}x = 0 \,\&\, p_{7,8}x \neq 0 \rightarrow P^{(6)}x = l(Vx) + a(Orgx - ORGx) + b(Dx - Sx))$$

where the latter may be termed "the law of market-value under the consideration of demand and supply" (Nowak 1980, 12).

The idealizations $p_3 = 0$, $p_4 = 0$, and $p_6 = 0$ entail that there is no competition between various branches of the economy. There is no flow of capital between them because the rate of profit, p', is the same in all spheres of production. But Marx removes the idealization $p_6 = 0$, that is, he assumes that $ORG_B x - ORG_E x \neq 0$. This enables him to account for the emergence of the average rate of profit in different branches of production and thereby for the transformation of market-values of commodities from separate branches of production into a unified price of production. For Marx the value of a commodity is the sum $v + c + m$, where $v + c$ is the advanced capital, symbolized as k. So, $L^{(6)}$ can be rewritten as follows

$$(x)(Cx \,\&\, p_{1-6}x = 0 \,\&\, p_{7,8}x \neq 0 \rightarrow P^{(6)}x = l(kx + mx) + a(ORGx - ORGx) + b(Dx - Sx))$$

Now with the removal of the idealization $p_6 = 0$ one arrives at the scientific law for the price of production

$$L^{(5)} : (x)(Cx \,\&\, p_{1-5}x = 0 \,\&\, p_{6-8}x \neq 0 \rightarrow P^{(5)}x = kx + kxp''x)$$

where p'' is the average rate of profit.

Subsequently, Nowak shows (1980, 16–19) that after the removal of $p_5 = 0$ it is possible to explain the profit of the merchant capital. By the removal of $p_4 = 0$, in turn, it is possible to derive the absolute rent and the abandonment of $p_3 = 0$ permits us to infer price deviations under the consideration of the fluctuations of surplus-value.

Proceeding from the above considerations, L. Nowak builds up both a general schema of a scientific law and a scientific explanation resting on that schema. Let me assume that the set U denotes "the universe of discourse" (1980, 28) and G denotes a propositional function called a realistic assumption, since it is fulfilled by any element of the universe U (e.g. "x is a commodity"). The next propositional function introduced is one termed "idealizational assumption" and denoted as $px = 0$, where "0 symbolizes the minimum value of the magnitude p" (1980, 28). It represents an abstraction from conditions labeled by Nowak as secondary factors and by me as modification conditions because they modify the phenomenal forms. The scientific idealizational law of the kth degree of idealization, according to Nowak, can be formulated as follows:

$$L^{(k)} : (x)(Gx \,\&\, p_{1-k}x = 0 \rightarrow E^{(k)}x = f_k(H_1x, \ldots, H_nx)) \tag{3.2}$$

The equation in the consequent expresses the dependence of the examined factor denoted by $E^{(k)}$, representing the phenomenal form in the kth degree of idealization, on the factors denoted by $H_1, \ldots, H_n$, which represent the principal factors or the essence of the investigated object. Such a structure, according to Nowak, is also appropriate for the scientific laws of physics. For example, the second law of motion in classical mechanics should have the structure (Nowak 1974, 29, 46):

$$L^{(2)} : (x)(Ox \,\&\, P(u) \,\&\, x \text{ is in } u \,\&\, Dx = 0 \,\&\, F_e u = 0 \rightarrow F_r^{(2)}x = mxax) \tag{3.3}$$

where O denotes a physical object, P the physical system, D all the dimensions of the physical object, F_e the force acting on the physical system, F_r the force acting on the physical object, m the object's mass and a its acceleration.

Further, on the basis of the structure of scientific laws having been explicated in such a way, Nowak constructs a model of scientific explanation of his own. Scientific explanation is understood by him as consisting primarily of a removal of the idealizational assumption $p_k = 0$ (that is, assuming the validity of $p_k \neq 0$), thereby necessitating the introduction of a "correction into the consequent" of $L^{(k)}$ "expressing the influence of the magnitude p_k upon the magnitude in question" $E^{(k)}$ (Nowak 1980, 29). Therefore, by the removal of the idealizational assumption and the introduction of the correction into $L^{(k)}$ the later is transformed into

$$L^{(k-1)} : (x)(Gx \,\&\, p_{1-(k-1)}x = 0 \,\&\, p_k x \neq 0 \rightarrow E^{(k-I)}x = f_{k-1}(H_1 x, \ldots, H_n x, p_k x))$$

where $f_{k-1}(H_1, \ldots, H_n, p_k) = g(f_k(H_1, \ldots, H_n, p_k), h(p_k))$ holds true. The function h "expresses the impact of the factor p_k on the investigated magnitude (correctional function), while g expresses the type of the impact; for example, the fact that it constitutes a strengthening or weakening of the dependence f_k; that is why g is called the directional function" (Nowak 1980, 29–30). One can gradually remove all the idealizations, thereby arriving at:

$$L^{(0)} : (x)(Gx \,\&\, p_{1-k}x \neq 0 \rightarrow E^{(0)}x = f_0(H_1 x, \ldots, H_n x, p_k x, \ldots, p_1 x))$$

Proceeding from $L^{(k)}$ to $L^{(k-1)}$ and down to $L^{(0)}$, by gradual removal of the antecedent idealizations and by taking into account the impact of the modification conditions, constitutes the method of scientific explanation by gradual concretization. In (1971; 1974) Nowak shows that physical theories have recourse to this method when it comes to scientific explanation. According to him, in the case for the second dynamic law of classical mechanics (3.3), for non-inertial systems it is necessary that two additional forces be taken into account: the force of inertia F_{in} and that of Coriolis F_c. (3.3) then changes into (Nowak 1974, 46)

$$L^{(1)} : (x)(Ox \,\&\, Pu \,\&\, x \text{ is in } u \,\&\, Dx = 0 \,\&\, F_e x \neq 0 \rightarrow F_r^{(1)}x = mxax + F_{in}x + F_c x)$$

Relying on his analysis of Marx's method in *Capital*, Nowak states that "I. Marx introduces idealizing assumptions, II. next he establishes the idealizational law, III. and he concretizes it gradually . . ." (1980, 31). Therefore, there obtains, *in Nowak's view*, a principal agreement between the type of scientific laws formulated in physics and those formulated by Marx in *Capital*. They appear to be scientific idealizational laws possessing the structure of (3.2). There ought to be an agreement in the employment of explanation methods; this has to be the method of gradual concretization. Although this type of scientific law and this method of scientific explanation, in some aspects, were anticipated by Galileo Galilei in physics, Nowak maintains that it was Marx who introduced them into political economy. Thus, Nowak insists, Marx is the Galileo of the social sciences (1980, 34–38).

3.2 WHAT HAS BEEN OMITTED FROM MARX

Before verifying Nowak's claim that Marx is the Galileo of the social sciences, let me try to make a list of the "facts" Nowak either completely neglected, or, in my view, misrepresented in his methodological analysis of Marx's economic writings.

When carrying on the reconstruction of the scientific law of value (1971), Nowak asserts the following: "since Marx has been keeping the law of value immutable throughout the two opening volumes of *Capital* until Chapter Eight of the third one, we shall enumerate assumptions on which the cited law is formulated in the portions of *Capital* immediately preceding the quoted eighth chapter" (1971, 30). Yet two objections can be raised against this claim. *First*, Marx in his "Sechstes Kapitel. Resultate des unmittelbaren Productionsprozesses," planned by him in the 1861–63 manuscripts (Marx 1980d, 1862), under the caption "Resultat des Produktionsprozesses," to follow the chapters on surplus-value and capitalist accumulation (i.e., it would in fact belong to the conclusion of the first volume of *Capital*) states that (1988b, 33; Marx and Engels 1975ff., Vol. 34, 363)

> [t]he *commodity* as it comes out of the capitalist production differs in its determinations from the commodity from which one proceeds as an element, presupposition of the capitalist production. We proceeded from the individual (einzelne) commodity as an independent article in which a certain amount (Quantum) of labour-time has been objectified . . .
> The commodity appears now as twice determined:
> 1) What is objectified in it . . . is a certain amount of socially necessary labor. But while in the case of the individual commodity it was entirely undetermined . . . from whom this objectified labor originated, etc., the *commodity* as the as the product of capital contains partly paid, partly unpaid labor . . .
> 2) The individual commodity appears not only materially as part of the total product of the capital, as an aliquot part of the lot produced by it. We have not at all in front of us the individual commodity, the individual product. Not singular commodities appear as the result of the process, but a *mass of commodities*, in which the value of the advanced capital + surplus-labor . . . has reproduced itself.

This in turn brings about a change of the law of value as it was initially given in Chapter 1 in the first volume of *Capital*. There it pertained to an individual commodity, but at its end we find out that (Marx 1988b, 33–34; Marx and Engels 1975ff., Vol. 34, 363)

> [t]he labor spent on an individual commodity . . . cannot to be calculated at all. It holds good only as an aliquot part of the spent on the mass of commodities, and ideally calculated, total labor. In the *determination of the price* of an individual commodity it appears only as an ideal part of the total product in which the capital reproduced itself.

This means that, contrary to Nowak's claim given above, the law of value at the end of the first volume of *Capital* cannot have the structure of (3.1), because in the latter value and price are related as yet only to isolated, individual commodities.

Second, a closer look at Part I (Chapter I, II, III) of *Capital* 's first volume, where such concepts as value, exchange-value, price, etc., are being treated, discloses the fact that the indicated portions of the work encapsulate neither the concept of capital nor that of surplus-value. What Marx describes in Part I is the relation between two producers-owners of commodities. If Nowak's claim about the law of value, introduced in chapter one in the first volume of *Capital*, were to hold, then in that chapter Marx must have already introduced the concept of variable capital, i.e., the concept of the value which the worker gets in exchange for the time during which he performs labor. But the opposite is true. Marx explicitly states here "[t]he reader must note, that we are not speaking here of the wages or value that the labourer gets for a given labour time . . . Wage is a category that, as yet, has no existence at the present state of our investigation" (1990, 37; 1967, Vol. 1, 44; Marx and Engels 1975ff., Vol. 35, 54). This, however, entails that out of the modification conditions p_1 through p_8 introduced by Nowak (1980) the following will have fallen out: p_3, since the concept of surplus-value has not been introduced as yet either, so there can be no reflection upon the organic composition of capital; factors p_4, p_6, p_8, since the concept of capital has not been introduced as yet either, so there can be no reflection upon the organic composition of capital; p_5, since there exists no class of capitalists as yet, which prevents reasoning about a social group of merchant capitalists as well as merchant profit. The scientific law of value, given in Nowak's understanding by (3.1), can by no means have such a form in Part I of volume I of *Capital* but, *at most*, only the following form:

$$L^{(3)} : (x)(Cx \,\&\, p_1x = 0 \,\&\, p_2x = 0 \,\&\, p_7x = 0 \rightarrow P^{(3)}x = f_3(Vx)) \qquad (3.4)$$

In this way we arrive at a startling conclusion. The scientific law of value in Part I of the first volume of *Capital* should involve three idealizations, while at the inception of the manuscript of the third volume, there must already be eight, i.e., their number has to increase as the discussion advances from the first to the manuscript of the third volume.

How can one account for this circumstance in terms of the Poznań School? Two possibilities immediately suggest themselves. One is that when progressing from the first volume to the manuscript of the third volume of *Capital*, Marx removes some of the hitherto cognized and, in the scientific law of value explicitly stated, modification conditions, i.e., from $p_k \neq 0, p_{k+1} \neq 0, \ldots, p_{k+r} \neq 0$ to $p_k = 0, p_{k+1} = 0, \ldots, p_{k+r} = 0$. But this would principally preclude the growth of knowledge about the commodity, this growth being achieved by a modification of the scientific law of value as given in the first volume of *Capital*, up to the form it assumes in the manuscript of the third volume. At the beginning of the first volume products are produced and sold as commodities, whereas in the manuscript of the third volume they are no longer "exchanged simply as *commodities*, but as *products of capitals*" (Marx 1992, 251; Marx 1967, Vol. 3, 175). But from Nowak's understanding of the structure of $L^{(k)}$ it follows that a purposeful removal of modification conditions expressed in a scientific law – i.e., putting them equal to zero – of necessity brings about a reduction of the relation in the consequent, the vanishing of its cognitive content. The ensuing schema makes it readily clear:

$$L^{(1)} : (x)(Gx \,\&\, p_1x = 0 \,\&\, p_2x \neq 0 \rightarrow E^{(1)}x = f_1(H_1x, \ldots, H_nx, p_2x))$$
$$L^{(2)} : (x)(Gx \,\&\, p_1x = 0 \,\&\, p_2x = 0 \rightarrow E^{(2)}x = f_2(H_1x, \ldots, H_nx))$$

The second possibility amounts to this. As he progresses from the first to the manuscript of the third volume of *Capital*, Marx discloses previously hidden idealizations about which he was unaware at the time he was writing the first volume. In other words, let it be given $L^{(1)}$, and suppose that the equation contained in it turns out to hold non-emptily only if a further idealization, say $p_3 = 0$, which was not yet known at the time of the formulation of $L^{(1)}$, is valid. So knowledge here unfolds from $L^{(1)}$ to $L^{(2)}$, where

$$L^{(2)} : (x)(Gx \,\&\, p_1x = 0 \,\&\, p_2x \neq 0 \,\&\, p_3x = 0 \rightarrow E^{(2)}x = f_1(H_1x, \ldots, H_nx, p_2x))$$

Yet this case, as well, fails to model accurately the passage from the first volume of *Capital* to the manuscript of its third volume, since the cognitive content of the functional dependence in the equation has remained unchanged, whereas we have shown that with Marx this cognitive content does grow.

The point of these two objections is that already in the first volume of *Capital* one has to reach the concept of capital proceeding from the scientific law of value, and this is attainable only through a passage from the law of value to the law of surplus-value. Although Nowak has pointed out that the explanation in the third volume of *Capital* is being carried out by the method of gradual concretization, nowhere (1971; 1974; 1980) has he accomplished a reconstruction of Marx's passage from the scientific law of value to the scientific law of surplus-value as that passage appears in the first volume of *Capital*. In chapter four I shall discuss his attempt at this in (1977).

Nowak also fails to reconstruct Marx's movement of thought in Part VIII of the first volume of *Capital* which allows him to understand the initial accumulation, or, in other worlds, to grasp the historical origins of the capitalist mode of production ("The So-Called Primitive Accumulation"), as well as to predict the decline of this mode of production ("Historical Tendency of Capitalist Accumulation"). Here Marx, through retrodiction in thought, reaches back to a society where surplus-value was not as yet produced, while by means of prediction he foresees the rise of a society no longer producing it.

And what about Nowak's approach to the scientific law of value as it is modified by gradual concretizations in thought? Given his $L^{(8)}$, the following can be demonstrated in the world wherein gradual concretization unfolds itself. At the outset, with all eight idealizations holding good, there is a world enclosed by a border (defined by the condition $E - I = 0$), and internally composed of the world of commodities, W_{c1}, and of golden money, W_m. The explanation begins with W_{c1} as a world containing only one element – a commodity which here "is to be considered as an average sample of its class" (Marx 1990, 33; 1967, Vol. 1, 39; Marx and Engels 1975ff., Vol. 35, 49), and let me label it as *a'*. The removal of the idealization $Org - ORG = 0$ causes the following change. W_{c1} comes to be inhabited not only by a commodity *a'* (produced by a capital with an average organic composition *ORG*), but also by a commodity a_1, produced by a capital with the organic composition *Org*. W_{c1} therefore changes into W_{c2}. By removing the idealization $ORG_B - ORG_E = 0$, one introduces a typical representative – the commodity *b* – produced by a capital with the organic composition ORG_E, while changing W_{c2} into W_{c3}. The latter falls into two subworlds:

the subworld-branch where commodities of the class *a* are produced (i.e. a' and a_1), and the subworld, covering the remaining branches, where *b* is being produced. The removal of $ORG_A - ORG_I = 0$ redivides the whole of W_{c3} into a subworld of commodities produced in agriculture and into a subworld of commodities produced in industry, creating thus a world W_{c4}. It becomes obvious that the entire process of gradual concretization here as a movement in thought is always bound up with a world of commodities, where these are products of capital, i.e., within each of the indicated worlds surplus-value is being produced. That is, as to its method, the movement in thought from W_{c1} to W_{c4} cannot be identical with the movement characterized by retrodiction and prediction which appear in Part VIII of the first volume of *Capital* – since at this juncture the very process of the production of surplus-value arises and fades away.

Let me now examine the state where W_{c1} and W_m occur. The value of commodity a' is expressed through a relation with an element from W_m as its price. But now it is possible to pose the following question. Is the world consisting of W_{c1} and W_m really the thought world Marx proceeds from in his account of the law of value in the first volume of *Capital*? I have insisted earlier that at its beginning Marx formulates this law not in the form of (3.1) but in the form of (3.4). Yet, as a matter of fact, Marx does not begin with the relation between W_{c1} and W_m, but only with a world *entirely identical* with the subworld W_{c1}. What first and foremost exists for Marx is *only* a' as a typical representative of the world of commodities, whereas the subworld W_m does not altogether exist. In the first volume of *Capital* (Chapter 1, Sections 1 and 2) Marx does not define (nor can he, since W_m does not exist) the price *P* through value *V*; instead he first defines value itself, the concept of money being arrived at only in the concluding portions of Section 3 of the same chapter ("The Form of Value or Exchange-Value").

If considered from this perspective, Nowak's theory as delineated in (1980) appears to have the following inconsistency. On the one hand, Nowak states that idealizations "are adopted to neutralize certain adventitious factors and thus to enable us to analyze relationships between the given phenomenon and its basic determinants only, that is those which affect most strongly the phenomenon under consideration" (1980, 96). In the case of the scientific law of value, as it has been reconstructed by Nowak, the above implies that the idealization method should enable one to express the relationship between price and value, since price is a phenomenal form of value. On the other hand, however, he argues that the idealization $D - S = 0$ in the scientific law of value "is adopted in order to eliminate the adventitious factor which is the fluctuation of demand and supply, and to concentrate on the essential factor, namely the value of the commodities, that is the time of socially necessary labour required to produce them" (1980, 96). Yet the introduction of this idealization entails something entirely different. The idealization should permit one to express not only the relationship between price and value, but ultimately also the relation between value and socially necessary labor. Nowak, however, in his reconstruction of the scientific law of value in (1971; 1980) explicates only the relationship of price and value, but never the relationship of value and socially necessary labor. For Nowak value is already given because its subject – the commodity – is already given. So for him it is possible neither to grasp the origin of value, nor, therefore, to relate it to anything but price.

The conclusion I arrive at, therefore, is as follows. Even if Nowak has reconstructed the scientific law of value containing the equation $P^{(8)} = l(V)$, yet since it involves $l(V)$ and therefore also V as a *definiens*, in accord with the system of knowledge Nowak finds at the beginning of the first volume of *Capital*, then there must occur a scientific law whose equation is a definition of value itself, i.e., where V is a *definiendum*. Nowak has not reconstructed such a form of the scientific law of value, where value itself is being defined, though he does implicitly assume the existence of this definition in the very beginning of the first volume of *Capital*. Therefore, in my view, his reconstruction of the consequent of the scientific law of value as it appears in the very beginning of the first volume of *Capital* is basically incomplete.

Let me try to rescue (3.1) and (3.4) by revising them so that they might withstand the criticism submitted above. I will replace the equation $P^{(3)} = l(V)$ by an equation expressing the relation between value and socially necessary labor, L, to produce a commodity x. Taking into account the fact that conditions p_1, p_2, and p_7 do not modify value, i.e., they can be left out from the scheme (3.4), the latter assumes the following form:

$$(x)(Cx \rightarrow Vx = f(Lx)) \tag{3.5}$$

Yet this attempt at revision is a failure because it does not express the inner connection between a commodity and its acquisition of a value equal to the amount of socially necessary labor to produce it. It does not exclude the case that there could exist a product created by socially necessary labor (having value as a result) that would still not be a commodity. This contradicts Marx's understanding of the relation of commodity to its value.[1] Already in the *Grundrisse* he claims that "the commodity *is* exchange-value" (1976, 120; Marx and Engels 1975ff., Vol. 28, 125). He claims also that "[t]he labour-time objectified in the use-values is . . . the substance which makes of them exchange-values and, therefore, also commodities . . . As exchange-value all commodities are only a certain mass of congealed labour-time" (1980b, 110; Marx and Engels 1975ff., Vol. 29, 272).

The problem Marx faced in his approach in (1980b), as well as in the first German edition of the first volume of *Capital* (1983), was that the term "exchange-value" referred both to the *internal* determinations of a commodity, its quantitative determinations (*Wertgrösse*), as well as to its *external*, manifest determinations, the quantitative determinations of the relative value (*Grösse des relativen Wertes*). Marx, therefore, by means of the manuscript *Ergänzungen und Veränderungen* (1987a) changed his terminology in such a way that, starting from the second German edition of the first volume of *Capital* (1987b), used the term "value" to refer to the internal determinations of a commodity, while restricting the usage of "exchange-value" to describe the external manifestation of value generated in the relation between commodities.

So, for example, the first subchapter, *The Commodity*, in the first German edition of the first volume of *Capital* (1983, 17) is relabeled in the second edition to read *The*

Two factors of the Commodity: Use-Value and Value (Substance of Value, Magnitude of Value) (1987b, 69). The caption *The Value-Form* from the addendum to the first volume of *Capital* was also subject to the same stylistic change (1983, 626); beginning with the second German edition, it was revised to read *The Value-Form or Exchange-Value* of subchapter 1.3 (1987b, 80). Thus, since the he second German edition of the first volume of *Capital,* (3.5) explicitly contradicts Marx's analyses of commodity. *A commodity is,* according to Marx, *a value-thing*. Nowhere and never could there exist a product of labor which would not be a commodity. The antecedent and consequent of (3.5) fail to match up with one another in another way as well. The antecedent expresses the fact that *commodity (a value-thing) and, therefore, also value, is already given*. But the consequent expresses only the *production* of value from socially necessary labor. According to Marx "labor as a process, *in actu*, is the substance and measure of value, not value" (1982a, 2099; Marx and Engels 1975ff., Vol. 34, 71). So L denotes here a certain *process* leading to the production of value and, therefore, also to the *production of commodity*.

My analysis thus far has been focusing on Nowak's construal of the law of value. As mentioned earlier, Nowak (1971; 1974; 1980) does not reconstruct the passage from the scientific law of value to the scientific law of surplus-value. But he (1974) makes an attempt at a reconstruction of the very law of surplus-value as it occurs in *Capital.* He holds that it contains the following idealizations: "the constant capital given by the producer equals zero," (1974, 72) which I shall denote as $c = 0$. "the merchant profit renounced by the producer in behalf of the commercial capitalist equals zero," (1974, 72) which I shall denote as $m_p = 0$. "the rate of interest paid by the production-capitalist to the banker equals zero (that is, the production-capitalist operates capital of his own and takes on no debt)," (1974, 72) which I shall denote as $i = 0$. "the property rental fees paid by the production-capitalist to the landowner equal zero (that is, the production-capitalist is the owner of the land, that is why there exists no group of land-owners)," (1974. 72–73) to be denoted as $r = 0$. If all these idealizations hold good, then Marx, according to Nowak, "has introduced the notion of surplus-value" according to which "surplus-value = value minus variable capital" (1974, 73). That is, if V denotes the overall value of a commodity and v the variable capital, then, given the four idealizations mentioned above, it should be possible to define surplus-value, m, as follows:

$$m = V - v \tag{3.6}$$

In my opinion Nowak's reconstruction fails to grasp Marx's understanding of surplus-value. Firstly, the concept of interest, merchant profit, and rent appear only in the manuscript of the third volume of *Capital,* while the concept of surplus-value is already discussed in the first volume, i.e., Marx defines surplus value independently of the other concepts. Secondly, in the first volume of *Capital* Marx defines surplus-value *before* taking into account "[t]he various factors of the labour-process" and how they "play different parts in forming the value of the product" (Marx 1990, 173; 1967, Vol. 1, 199; Marx and Engels 1975ff., Vol. 35, 209). That is, he first (in Chapter VII) introduces the concept of surplus-value, and only then (in Chapter VIII) conceptually distinguishes between the constant and variable components of capital. Thirdly, even when taking into account Marx's understanding of surplus-value *after* the introduction of the concept of constant capital, it turns out that surplus-value emerges independently of the quantitative characteristic of the factor denoted by c. If one takes into consideration the fact that the value of the constant capital is being transferred through the labor process into the value produced, then this new value contains c (if not the whole of it, then at least its part), thereby the invested capital does not equal v, but $v + c$. Hence (3.6) may be restated ($V' = V + c$ with regard to (3.6)) as

$$m' = V' - (v + c) \tag{3.6'}$$

and so the fact that surplus-value is contained in V' is completely independent of the quantitative determination of the constant capital. Fourthly, (3.6) makes it apparent that Nowak's understanding of surplus-value is *circular*. V is here, from the point of view of its understanding in Chapter VII of the first volume of *Capital*, comprehended as containing both m and v, i.e., $v + m$. So Nowak's reconstruction expressed in (3.6) really means nothing else than the triviality

$$m = (m + v) - v \tag{3.6''}$$

Fifthly, in his understanding of surplus-value Nowak assumes, as he did when considering the law of value, that the commodity which contains objectified surplus-labor has already come into being. He does not reconstruct Marx's reflections on the relation of surplus-value, *SV*, to its origin, to the process of surplus-labor, *SL*, in which the former originates. Marx's law of surplus-value from Chapter VII of the first volume of *Capital* contains, therefore, the equation $SV = k(SL)$, where k denotes a function, but the antecedent conditions of this equation have yet to be determined; it is already apparent that they are different from the idealizing assumptions $c = 0$, $i = 0$, $r = 0$, and $m_p = 0$.

Our analysis of Nowak's reconstruction of Marx's *Capital* arrives at a startling conclusion. Nowak "enters" into its third volume and everywhere where Marx operates with certain modification conditions (i.e. they are here non-zero), he makes them *retroactively* equal to zero, and thereby tries to reduce the phenomenal forms (again *retroactively*) to their essence, the essence from which they were previously derived by Marx. But he fails in his attempts to comprehend the law of value and law of surplus-value as a result of this act of reduction. Stated in even stronger terms, Nowak's reconstruction fails even if, by "entering" into the world of phenomena of the third volume of *Capital*, he is in the favorable position of having, so to speak, an *a priori* knowledge of all the concepts introduced in its first and second volume. This critique should make it readily clear that if one accepts the first volume together with the manuscripts of the second and third volume of *Capital* as an explanatory structure, then one must *initially* reconstruct the scientific laws and methods contained in its first volume, and only *then* move on to reconstruct the methods of scientific explanation by means of which one progresses to its third volume.

A complete reconstruction of the scientific law of value and surplus-value from the first volume of *Capital* requires the introduction of a *concept of conditions different from the concept of modification conditions*, with the former concept being related somehow to the *very concept of essence*. But it is exactly this concept that is missing in Nowak's works; he conceptually reconstructs only conditions related to the phenomenal forms $E^{(k-1)}, \ldots, E^{(0)}$.

3.3 THE PURE IDEALIZATIONAL TYPE OF SCIENTIFIC LAW

The previous subchapter may have given the impression that Nowak's understanding of idealizational laws and of methods of scientific explanation based on them is at odds only with Marx's *Capital*, but not with the structure of scientific laws in physics. This impression is wrong. Let me analyze the equation appearing in the consequent of (3.2). Nowak maintains that "[a]ccording to the proper scientific method, an investigator should separate the principal and secondary factors for a given *phenomenon* (emphasized by me-I.H.) and abstract from them establishing the law connecting the phenomenon with the principal determinants of it" (1980, 37). *Ultima ratio* for him is the equation relating the phenomenal form $E^{(k)}$ with the principal factor H, and which, he believes, has the form of

$$E^{(k)} = f_k(H) \tag{3.7}$$

In my view, he commits here a *fundamental mistake*. According to Nowak, the formula (3.7) should represent the schema of equations that outline both the scientific laws of *Capital* and of physics. So let me compare the equation $F_r^{(2)} = ma$ given in (3.3) with the equation $P^{(8)} = l(V)$. The latter expresses not only the fact that value is the basis for the phenomenal form of price in the eighth degree of idealization, but, apart from this, that the very concept of price is being derived by Marx from the concept of value, and this is so because in *Capital*, before the cited derivation has been carried out, the very concept of value acquires its definition through the concept of socially necessary labor, L. Since Marx first applies the definition $V = k(L)$, he will afterwards have arrived at the equation $P^{(8)} = l(V)$, i.e., at a definition of price through value. With Marx, therefore, the equation (3.7) does indeed hold good. According to Nowak, (3.7) applies also to the equation $F_r^{(2)} = ma$ for the second dynamic law of classical mechanics. In his view, with respect to the two idealizations given in (3.3), the second dynamic law of classical mechanics can be restated as $F_r = f_2(m,a)$. This would entail that acceleration and mass are the principal factors, the force being their manifestation. According to the postulates of classical mechanics, however, this is a principal error. It is force that is the *cause-ground,* and its *effect – the grounded* is acceleration of the body with a given mass. A physically adequate representation would be $a^{(2)} = F_r/m$. But the point is that classical mechanics is a science that only deals with the effects of the forces, not with their origin. Therefore, even if equation $a^{(2)} = F_r/m$ holds for the relation of essence and phenomena (its structure corresponding to those of (3.7)), nevertheless, in terms of the definition of classical mechanics, the formula $F_r := ma^{(2)}$ holds (":=" stands for definition). In classical mechanics the general schema, contrary to the schema in Marx's *Capital*, is as follows:

$$f_1(H) := E^{(k)} \tag{3.8}$$

The essence-ground-cause *H has to be* defined through the phenomenal form-effect $E^{(k)}$.

Someone could object to my distinction between (3.7) and (3.8) by suggesting that we should flip the sides of the equation (3.8) to bring it into accord with the equation (3.7). But such a reformulation cannot work because in the equation $E^{(k)} = f_1(H)$, obtained by reversing the sides of the equation, the essence *H* still cannot be expressed independently from the phenomenal form $E^{(k)}$. One can here replace *H* only by $f_1^*(E^{(k)})$, where f_1^* denotes a function inverse to that denoted by f_1. But then the substitution for *H* in $E^{(k)} = f_1(H)$ ends up in the triviality $E^{(k)} = E^{(k)}$. That is, the flipping of the sides in equation (3.8) is only a tautological reformulation, a sheer *formality*.

The reconstruction of equations with the structure corresponding to (3.8) enables me also to bring the critique to Mackie's claim – that it is possible to purify the mixed laws of classical mechanics completely because they contain full knowledge of a Newtonian mechanical system – to its final form. Mackie never realized that accidentality resides even inside the basic law (2.8). Since the origin of force is in classical mechanics completely unknown, $a = f/m$ does not hold by definition. Classical mechanics does not explain why force necessarily manifests itself as acceleration.

By comparing (3.8) with (3.7) I am led to a conceptual distinction never accomplished by Nowak. Phenomenal forms obtained, cognized before the cognition of their essence, and by means of which it is possible to define the latter, I will label *forms of appearance of the essence*. Phenomenal forms obtained, cognized *after*, and inferred *from* the cognized essence, I will label *forms of manifestation of essence*. This distinction corresponds to the distinction made by Marx by means of the German terms *Schein* and *Erscheinung*.[2] Wherever Marx derives phenomena on the basis of the already cognized essence, say the derivation of the exchange-value from value in chapter one in the first volume of *Capital*, he uses the term "form of manifestation" (1990, 32, 48; 1967, Vol. 1, 38, 56; Marx and Engels 1975ff., Vol. 35, 48, 67). Very often when Marx mentions the case when there is knowledge of the phenomena but as yet no knowledge about their genesis from and relation to the essence, he uses the term "appearance." So, for example, he states (1979, 1450; Marx and Engels 1975ff., Vol. 32, 449)

> [t]he forms of revenue and the sources of revenue express the relation of the capitalist production in the *most fetish-like form*. It is their existence (Dasein) as it appears on the surface, separated from the concealed connections and the intermediate links.

Now it is possible to reformulate (3.3) and thus bring it into accord with the second dynamic law of classical mechanics. One obtains

$$L^{(k)} : (x)(Nx \,\&\, Cmod_{1-k}x = d_{1-k} \rightarrow f_1(Hx) = E^{(k)}x) \tag{3.9}$$

where *N* denotes the natural kind for instances of which the law is formulated; $f_1(H)$ denotes a certain function of the cause-ground-essence; $E^{(k)}$ is the appearance-effect in the *k*th degree of idealization; $Cmod_{1-k}$ denotes a conjunction of 1 to *k* ($k \geq 1$) modification conditions (secondary factors); d_{1-k} means a conjunction of 1 to *k* zeros, constants or intervals. $Cmod_{1-k} = d_{1-k}$ is a shorthand for the conjunction $Cmod_1 = d_1$ & . . . & $Cmod_k = d_k$ and represents a conjunction of 1 to *k* abstractions from 1 to *k* modification conditions. This type of idealizational law was anticipated by Hegel in his *Phenomenology of Mind*, where he used the term "pure law" for it (Hegel 1967, 293).[3] Therefore, I will be referring to scientific laws with the structure of (3.9) as the *pure idealizational type of scientific law of essence* in the *k*th degree of idealization. This category of scientific laws must not be confused with the laws contained in the third volume of *Capital*. They differ from each other, at least, by the structure of their respective consequents. The structure of equations given in the third volume of *Capital* corresponds to (3.7), not to (3.8). In chapter six I will compare the equation (3.8) with the equation (3.7) by testing them in relation to the category of measure. I will show that while the former expresses knowledge of the apparent (external) measure, the latter expresses already knowledge of the manifestation of the immanent measure.

3.4 $L^{(k)}$, SCIENTIFIC EXPLANATION AND THE D-N MODEL

In my view, a scientific law with the structure of (3.9) represents a more developed type of scientific law than the empirical law of coexistence and the empirical causal (Humean) law I analyzed in chapter one. The articulation of relations obtaining among phenomena does not qualify any longer as the sole aim of scientific knowledge expressed in a law with the structure corresponding to (3.9). The aim already is to reduce effects-appearances to their ground-cause-essence, i.e., to the mechanism which generates them, the ground from which the effects ought then to be derived. Such a type of scientific law should contemporaneously permit, contrary to the empirical causal type of scientific law, discrimination between the case where the equation of a scientific law conveys an ontic bond between the effect's genesis and its cause and the case of an epistemic approach, where the equation is only modified in order to identify the cause by the effect. In this perspective, the pure idealizational law of essence also can be termed the *theoretical (non-Humean) causal type of scientific law*.

By what types of objects are laws of the type $L^{(k)}$ fulfilled? In Nowak's view, because they contain terms (mass point, ideal gas, etc.) which are "defined as empty in [the] empirical domain, as denoting no real object," (1980, 65) they "non-emptily hold for ideal types and do not hold for real objects" (1971, 41). Contrary to this, in my view, a law of the $L^{(k)}$ type is formulated so as to be non-vacuously fulfilled by real objects. But here the term "real" is understood in a way different from Nowak. I do not identify real objects only with their forms of appearance. A scientific law which has the character of a pure idealizational law of essence cannot be fulfilled by the apparent forms of the objects simply because when one formulates a law of this type, he *abstracts* from these forms, converts them to their ground-cause-essence. In my view, when one formulates a law of the type $L^{(k)}$ type, the goal is that it should be non-vacuously fulfilled by real objects of a certain natural kind which have been reduced by thinking (in thought) to the essence of that natural kind. In the following two subchapters I will evaluate the success of this ambition.

A scientific law with the structure expressed by (3.9) provides a foundation for two types of explanatory procedures. The first of the two is a two-way movement of thought. On the one hand, it is a subsumption of the individual phenomenon $E^{o}a$ to be explained under a pure idealizational law of essence. What has to be emphasized here is that $E^{o}a$ is *not subsumed under a covering law but only under the covering natural kind and the latter's (formal) ground*, where both are expressed in the scientific law of the $L^{(k)}$ type. On the other hand, it is a piecemeal removal of antecedent idealizations accompanied by a gradual consideration of the causal effects of the modification conditions upon the phenomenal form $E^{(k)}$ in the consequent. I label that second aspect of explanatory procedure, in terminological accordance with Nowak, *gradual concretization*. I will return shortly to this point and show how my reconstruction of this procedure differs from his. It is a thought procedure for the approximation of the scientific law to the modification conditions of the phenomenon to be explained. Through it scientific knowledge first evolves from the law of the type $L^{(k)}$ to a law with the structure

$$L^{(k-1)}: (x)(Nx \,\&\, Cmod_{1-(k-1)}x = d_{1-(k-1)} \,\&\, Cmod_k x \neq d_k \rightarrow E^{(k-1)}x = E^{(k)}x + g_k(Cmod_k x)) \qquad (3.10)$$

$Cmod_{1-(k-1)} = d_{1-(k-1)}$ means that the first idealization through $(k-1)$ idealization is still valid; $Cmod_k \neq d_k$ denotes that the kth idealization does not hold any more; g_k is a function which conveys the causal influence of $Cmod_k$ upon $E^{(k)}$. By having removed the last idealization, one will arrive at the law with the structure

$$L^{(0)}: (x)(Nx \,\&\, Cmod_{1-k}x \neq d_{1-k} \rightarrow E^{(0)}x = E^{(k)}x + g_k(Cmod_k x) + \ldots + g_1(Cmod_1 x)) \qquad (3.11)$$

Gradual concretization is an explanatory procedure by means of which one explains theoretical causal laws from theoretical causal laws. So, the model of scientific explanation by gradual concretization represents a reconstruction of a type of scientific explanation never accomplished in the framework of the D-N model.

Consequently, after arriving at a scientific law of the type $L^{(0)}$, one can take into account the determinations of a specified individual object *a* and, by introducing those determinations into a law with the structure of (3.11), arrive at the explanation of the individual phenomenon $E^{o}a$.

My understanding of this method differs from that of Nowak by the fact that in my view the corrections in the consequent can be made only with respect to the form of appearance $E^{(k)}$, whereas he claims that they are made with respect to $f_k(H)$. But this is again equal to the false claim that, for example, classical mechanics is able, prior to the explanation of the manifestations of force (movement of the heaven bodies, the free fall, the pendulum), to define force independently of these phenomena. As an example of this let us take the case of the derivation of the law of the free fall from Newton's second law. Knowing the force acting on a body, say, on earth, from its effect on this body – acceleration g – we can derive the path covered by that body under the influence of that force. If we suppose that on earth the force acts on a falling body, which we initially regard as a mass-point, only in the direction of the y-coordinate, then we proceed from the following three differential equations: $F_x = md^2x/dt^2 = 0$; $F_y = md^2y/dt^2 = -mg$; $F_z = md^2z/dt^2 = 0$. By integrating them twice and by a suitable choice of the beginning of the coordinate system we derive the law of free fall in (1.7). The cornerstone of the whole derivation of the *effect as manifestation* of F – the covered path – is the *prior* knowledge of the *effect as appearance* – g – of the force acting on the body in the y-coordinate. Because in Newton's second law we cannot characterize force F independently of its effect-appearance – acceleration g (on earth) – but only by means of $F = mg$, we can neither substitute in the law of free fall (1.7), or in any law derived from it by gradual concretization, the term g by the term F/m. Such a substitution would end up in the trivial substitution of g for g.

The second explanatory move deriving from a scientific law of type $L^{(k)}$ rests on a derivation of the effects as manifestations from the cause-essence-ground, grasped by means of the relation $f_1(H) = E^{(k)}$. For this purpose we construct certain idealized objects and then derive the concrete law for those effects as shown, e.g., for the free-falling mass point under the impact of the force F acting in the direction of the y-coordinate. Similarly, the law for the simple pendulum is being inferred from the findings of kinematics and dynamics as well as from the theory of rigid bodies on the basis of which one shapes a thought object of the compound pendulum. The latter is described by means of a differential equation whose modification will yield the expression for the pendulum's period of oscillation. If one then converts the compound pendulum into a simple one, by assuming that the whole mass is contained in the bob, he or she will finally arrive at the equation $T^{(8)} = 2\pi\sqrt{l/g}$.

In the first chapter I tried to show that empirical laws of coexistence and succession are formulated at the level of empirical generalizations, and that explanations based on them purport to explain a certain individual phenomenon. With regard to a law of the $L^{(k)}$ type, those types of law come out as laws of appearance,[4] while the individual phenomenon $E^{o}a$ is the individual appearance of the ground H. At the level of theory formation, which succeeds the level of empirical generalizations, laws with the structure corresponding to (3.11) are formulated. Relying on these laws as a foundation, one can accomplish explanatory procedures aimed at the derivation of laws of the $L^{(0)}$ type, hence also at a derivation of $E^{(0)}a$ from the ground H. Regarding $L^{(k)}$, laws of the type $L^{(k-1)}, \ldots, L^{(0)}$ turn out to be the laws of the manifestation of essence, while $E^{(0)}a$, inferred from $L^{(0)}$, presents the individual manifestation of essence.

Now it is possible finally to assess Hempel's D-N model of scientific explanation. Firstly, Hempel's inability to reconstruct modification conditions expressed in scientific laws, as well as his understanding of the meaning and place of idealizations in them, has negatively affected his reconstruction of scientific explanation in his article (1948). The D-N model reconstructs scientific explanation only as a *one-way* process of *subsumption* of an individual phenomenon under a covering scientific law, *while completely neglecting the thought process going into the opposite direction*: the concretization of the scientific law to the modification conditions of the phenomenon to be explained.

Secondly, the D-N model is an explication of explanatory moves resting exclusively on scientific empirical laws and permitting only a grasp of a certain singular phenomenon-event.[5]

Thirdly, I would insist that Hempel is mistaken when he states that explanations of "empirical laws from the level of empirical generalization" based on theoretical laws and the formation of theoretical principles derived from the level of theory formations have the character of a non-causal D-N explanation (Hempel 1965a, 352). At the transition from Newtonian laws to the law of the simple pendulum or free fall, there occurs a *transposition* of the description of the cause from the former to the latter. The concept of force is explicitly embraced as $g^{(2)}$ in the relation $s^{(9)} = g^{(2)}t^2/2$ and $T^{(9)} = 2\pi\sqrt{l/g^{(2)}}$.[6] If the indicated explanation is said to have a non-causal character, it does so only in the sense that the concept of causation it employs is inconsistent with the Humean construal of causation. Nor does the derivation of the law of free fall and of the simple pendulum have the character of a deductive argument. Neither one is a derivation of cut-and-dried knowledge in the form of consequences from given premises, but a heuristic process creating new knowledge of the causal genesis of manifestations. It involves a thought construction of idealized objects as well as a deliberate introduction of additional idealizations. The heuristic nature of this process is also evident in the fact that the explanatory move seeks to convert a scientific phenomenological law (law of appearance) into a scientific law of manifestation as well as to transform knowledge about an individual appearance $E^{o}a$ into an individual manifestation $E^{(0)}a$, by which one currently arrives at a growth of knowledge.

Fourthly, Hempel has committed two principal errors in his reconstruction of explanatory procedures in physics. He has claimed that the D-N model, together with the concept of "scientific law" that it involves, is the *ultima ratio* of physics. He was unaware of the fact that the model conforms only to pre-Newtonian physics, i.e., only to the level of empirical generalizations. With the advent of Newtonian physics, mechanics passed over to the formulation of a more advanced type of scientific law, as well as a more advanced method of scientific explanation, i.e., it escalated to the level of theory formation. Hempel has been blinded by his own absolutization of the D-N model as the only model of scientific explanation containing universal laws and for that reason has neglected to conduct a detailed examination of explanatory moves at the level of theory formation.

3.5 DOES THE PURE IDEALIZATIONAL TYPE OF SCIENTIFIC LAW REPRESENT THE *ULTIMA RATIO* OF SCIENCE?

The ambition of a scientific law of the type $L^{(k)}$ is, I have claimed, to reduce phenomenal (apparent) forms-effects to their ground-essence-cause; from their ground-essence-cause, they then ought to be theoretically derived as forms of the manifestation of essence. But a closer look at the structure of $L^{(k)}$ and at the method of explanation by gradual concretization makes it readily clear that this ambition cannot be fully satisfied. Firstly, the reduction of appearances-effects to their ground-cause-essence is *incomplete*, because in the equation $f_1(H) = E^{(k)}$ the essence is still defined through a form of appearance (even if in the kth degree of idealization); a complete fulfillment of the ambition amounts to a definition of the essence independently from appearances (and prior to the derivation of the manifestations). Secondly, the equation $f_1(H) = E^{(k)}$ involves an inherent contradiction between the genesis of the cognition of the cause-effect relation, expressed by the concepts of the *reason* and the *reasoned (by)*, and the real genesis of the cause-effect relation, expressed by the concepts of *ground* and *grounded (by)*. On the one hand, since the effects-appearances are cognized prior to the recognition of their ground-cause-essence, and the essence in itself is not as yet cognized, the latter has to be defined – *reasoned by* means of them. But on the other hand, this ground ontically gives rise to these effects, it *grounds* them. The initial goal of the formulation of a scientific law of the type $L^{(k)}$: that the *essence-ground-cause* will be the *reason*, and the *effects-appearances* will be the *reasoned*, is not attained because the result is the opposite one. The effects are the reasons and the cause is the reasoned (by them).

This fact is also evident in the process of gradual concretization which has its beginning in a law of the type $L^{(k)}$. Contrary to initial scientific ambition, the manifestations $E^{(k-1)}, \ldots, E^{(0)}$ are not explained on the basis of the essence H, but on the basis of the appearance $E^{(k)}$. Therefore the aim to transform a scientific law of the appearance of essence into a scientific law of the manifestation of essence cannot be attained completely. As I have shown in the case of the second dynamic law of classical mechanics and the law of free fall, any attempt to replace $E^{(k)}$ in the process of scientific explanation by $f_1(H)$ is doomed because it only amounts to a formal reversal of the sides in the equation $f_1(H) = E^{(k)}$. Therefore the *ground-essence-cause* H, as it is cognized in a scientific law of the type $L^{(k)}$, is as yet only a *formal ground*.

The term "formal ground" was introduced by Hegel in (1923, T. 2, 76–82; 1969, 456–461). After showing that classical mechanics grasps the ground of all mechanical phenomena only as a formal ground and not as a real ground, he in effect deprived classical mechanics of the status of true scientificity (on Hegel's relation to Newton see (Ihmig 1993)). In my view such an approach is too radical. Here, as in subchapter 2.7, my expressed view is that scientific knowledge as it develops through time is always at a particular stage of this development, and at certain stages the determinations of the objects under scrutiny need not be immediately cognized completely. Therefore, scientific knowledge with a structure corresponding to (3.9) should not be deprived of the status of a scientific law; one has simply to bear in mind that a particular law is not the *ultima ratio* of science; that science has to develop a more advanced type of a scientific law as compared to that of the $L^{(k)}$ type. In subchapter 6.1 I will deal with the formal ground and the real ground as they relate to the category of measure. Here I will show that knowledge based on the scientific law with the structure of (3.9) yields at least the information about the quantity (size) of the cause-essence-ground.

The conclusions to be drawn from this study of the reconstruction of the pure idealizational type of scientific law of essence follow here. Firstly, scholarship related to the explication of the pure idealizational type of the scientific law of essence succeeded in providing a concept of scientific law more developed than the concept of empirical law. Secondly, it succeeded in reconstructing a more developed method of scientific explanation as compared to the method characterized by deductive subsumption. Thirdly, it is still necessary to reconstruct, with respect to $L^{(k)}$, a more developed type of scientific law, where the essence would be the reason and the manifestations would be the reasoned (reasoned from the essence); in short, a law which would yield the concept of a real ground. Fourthly, it is necessary to reconstruct the explanatory methods based upon such a type of a scientific law. Fifthly, the success of these two reconstructions should be measured by their capability to a) grasp the structure of the scientific law of value and surplus-value as they appear in the first volume of *Capital*; b) reconstruct the basis upon which Marx makes his retrodiction and prediction on surplus-value in Chapter VIII of the first volume of *Capital*; and c) reconstruct the method by means of which Marx explains the law of surplus-value on the basis of the law of value in the same volume of *Capital*.

I will make an attempt at such a reconstruction in part three of this book. Prior to this I am going to reconstruct the ontological foundations of the pure idealizational law of essence.

CHAPTER 4

ONTOLOGICAL FOUNDATIONS

4.1 THE CONDITIONS OF A LAW AND NATURAL NECESSITY

Now it is possible conceptually to distinguish singular conditions from modification conditions. The fact that the concept "initial conditions of a singular event" has never, to the best of my knowledge, been explicated, is rooted in the absence of a reconstruction of a more developed concept of conditions of a law as compared to the concept "singular conditions." Let me make an attempt at such a reconstruction, utilizing Popper's ideas about the law of gravitation, mentioned earlier in subchapter 2.3.

Let us imagine a universe as a set of those possible worlds, each of which is inhabited by one pair of individuals with a non-zero mass. In Popper's understanding the law of gravitation should have the structure

$$(x)(y)(Ox \,\&\, Oy \rightarrow Fxy = kmxm'y/rxy^2) \qquad (4.1)$$

where O denotes objects with a non-zero mass; F denotes the force of interaction between these two objects; m and m' their respective masses; and r their distance. In my opinion, at this point Popper failed to take into account certain favorable (ideal) conditions, exactly in the same way that he neglected their reconstruction in the moa case. The equation in (4.1) holds only if one presupposes that the volume of the interacting bodies is equal to zero ($V = 0$) and only if the physical systems, into which these bodies are placed, are not subject to the impact of an external force ($F_e = 0$). So, (4.1) has to be restated as follows:

$$(x)(y)(Ox \,\&\, Oy \,\&\, Vx = 0 \,\&\, Vy = 0 \,\&\, F_e x = 0 \,\&\, F_e y = 0 \rightarrow Fxy = kmxm'y/rxy^2)$$

These ideal conditions have the character of an abstraction from the type of condition which characterizes a pair not just in one possible world, but in more than one possible world. Therefore, it is possible to predicate their instantiation of more than just one possible world, i.e., their applicability for individuals which exist in different possible worlds and which are denoted by individual variables. Modification conditions which co-generate (together with the essence) the manifestations $E^{(k-1)}$,. . . , $E^{(0)}$ can be predicated for more than one possible world. Contrary to this, I understand a *singular condition* as the type of condition which can be *predicated just for one possible world and its "inhabitants" denoted by proper names.*

The claim that a modification condition, contrary to a singular condition, can be predicated for more than one possible world *suffices* conceptually to determine a singular condition. But, it is completely insufficient for determining the very concept of modification conditions. First, it is not clear whether they can be predicated for all, or only for some, possible worlds inhabited by instances of the natural kind N. Therefore, second, it is also unclear whether they hold necessarily or only accidentally in these worlds. For a complete determination of the concept of modification condition one has to provide, with respect to it, a more developed concept of conditions of a law.

The conceptual differentiation between modification conditions and singular conditions implies a view of scientific explanation which is different from the Popperian view. According to Popper, it should be possible, on the basis of scientific laws and singular conditions, to explain a singular event. But, because he understands scientific explanation primarily as particularization – i.e., as substitution of individual values for variables appearing in the scientific law – one can arrive at an explanation of the singular event only insofar as it corresponds to that kind of event which appears in the scientific empirical law serving as the basis of explanation. This was already realized by Hempel who, working from the D-N model, stated that "[w]hat might in fact be explained is rather the *occurrence of a particular instance of a given kind of event*" (1965a, 423). To explain a certain singular event in its individual specificity amounts to an inclusion of conditions into the explanation process which can be predicated only for individuals existing in one possible world and denoted by proper names.

The conceptual differentiation between modification and singular conditions requires an alteration of Popper's definition *(N^0)* of natural necessity. I label both modification conditions and singular conditions as *secondary conditions*, reserving the term "primary conditions" for the conditions of the existence of the very essence, and provide the following definition[1]:

(N*) A statement may be said to be naturally or physically necessary if, and only if, it is derivable from a statement function which is satisfied in all worlds which differ from ours, if at all, only with respect to secondary conditions.

(N)* provides a further substantiation of the idea expressed in subchapter 3.4, the idea that only the inclusion of modification conditions into scientific laws makes it possible to utilize the former in explanations of scientific laws. They are conditions of the modifications of *universal* forms of manifestation expressed in scientific laws, and therefore the derivation of other universal forms of manifestation is equivalent to the derivation of a scientific law from a scientific law. But *(N*)* is still unable to escape the Nerlich-Suchting charge of circularity. The central problem remains the same: How to determine that set of possible worlds in which, even if they differ from each other (at most) with respect to secondary conditions, the law is non-trivially (non-vacuously) true, without getting trapped in a vicious circle? The concept "natural kind" appears to offer a possible solution.

4.2 A NATURAL KIND AND ITS ESSENCE

The issue of how to determine the set of all possible worlds in which a certain law should hold without being trapped in a vicious circle can, at least in my view, be solved if one reconstructs the starting points of Newton's *Principia* and that of Marx's first volume of *Capital*. In both of them the first step of the thought-movement is the determination of the *universe of discourse*. Newton starts with Definition I, where he claims that "[t]he quantity of matter is the measure of the same, arising from its density and bulk conjointly" (1946, 1; 1687, [1]). In this step Newton determines the character of entities which he will deal with: objects which are characterized by a *quantity of matter* (*quantitas materiae*) or *body* (*corpus*), or by what we today call *mass*.

Marx determines in a similar way the framework of his economical treatment, stating that "[t]he wealth of those societies in which the capitalist mode of production prevails, presents itself as an 'immense accumulation of commodities"' (1990, 29; 1967, Vol. 1, 35; Marx and Engels 1975ff., Vol. 35, 45). Commodities are chosen by him as the "universal elementary form of wealth," (1988b, 28; Marx and Engels 1975ff., Vol. 34, 369) from which he then proceeds further.

Newton, after determining the universe of entities, goes on by determining the phenomenal form which characterizes them. He chooses as such explicitly the *quantity of motion* (*quantitas motus*) and claims in Definition II "[t]he quantity of motion is the measure of the same arising from velocity and quantity of matter conjointly" (1946, 1; 1687, [2]). By means of this phenomenal form, as well as by the implicitly supposed phenomenal form of *acceleration*, he then determines the essence-ground-cause, the *centripetal force* (*vis centripetae*) in Definition VII and VIII as follows (1946, 4; 1687, B2)

> [t]he accelerative quantity of a centripetal force is the measure of the same, proportional to the velocity it generates in a given time. . . . The motive quantity of a centripetal force is a measure of the same, proportional to the motion it generates in a given time.

Only after Newton moved from the concept of quantity of matter, via the concept of quantity of motion (and of acceleration) to the determination of the centripetal force, does he finally arrive at his three laws of motion (Lex I, II, III).

Marx similarly, after determining the universe of entities, goes on by determining the phenomena which characterize them: the *use-value* and *exchange-value*. Marx then uses one of them – exchange-value – to determine the ground-cause-essence of those entities, *value*. Finally, Marx provides the law of value by relating value to its substance – labor.

What is readily seen in both of these cases is that one faces here a multistage cognitive movement: *the natural kind*; *the phenomena of the entities of this kind*; *the cause-essence-ground of these phenomena*; and finally *scientific law*; i.e., the category of natural kind is prior to that of scientific law. In Chapter 6 I will deal with the relation of the category of natural kind with that of scientific law on the basis of the category of measure and I will compare on this basis also Newton's thought-movement from mass to force with that of Marx from commodity to value. In Chapter 8, finally, I will show how the categorial sequence *natural kind* → *scientific law* is reversed into the categorial sequence *scientific law* → *natural kind*.

4.3 L. NOWAK'S ONTOLOGY AND ITS REFORMULATION FOR $L^{(k)}$

Nowak (1977; 1980) undertakes an attempt at formulating an ontological foundation for his understanding of the concept "scientific law" and for his understanding of scientific explanation. He introduces the set P_E which stands for the "set of all those factor magnitudes that have influence upon" E. "We shall term the set" P_E "the space of essential magnitudes for" E (1980, 97). He discriminates within his ontology between two types of essential magnitudes: the so-called *principal factors* for E, designated as $H_1, \ldots, H_n$, and secondary factors, symbolically $p_1, \ldots, p_k$. All these factors make up a certain hierarchy, which he calls the *essential structure*, and I call the *structure* S_E of the factor E. For one principal factor, the hierarchy can be represented as follows

$$\begin{array}{ll} (k) & H \\ (k-1) & H, p_k \\ \vdots & \vdots \\ (1) & H, p_k, \ldots, p_2 \\ (0) & H, p_k, \ldots, p_2, p_1 \end{array}$$

It consists of significance levels $(k), (k-1), \ldots, (1), (0)$. The first of them stands for the internal level, the last for the surface level. Nowak then unfolds his ontology through the following concepts. Let us have a structure of the phenomenon E so that the following holds

$$\begin{array}{ll} S_E & G \\ & G, p_1 \\ & G, p_1, p_2 \end{array} \tag{4.2}$$

Here G denotes the principal factor (the essence), p_1 and p_2 denote the secondary factors (the modification conditions). The relation of E to these factors is given by

$$\begin{aligned} E &= f(G) \\ E &= g(G, p_1) \\ E &- h(G, p_1, p_2) \end{aligned} \tag{4.3}$$

where the dependence f, g, and h make up the nomological structure N_E of the factor E. By means of N_E and S_E Nowak defines the concept of the *object category*, which for E in a universe U means the following ordered triple:

$$K_E = (U, S_E, N_E)$$

The concept of the object category, specifically its motion, is the focus of Nowak's investigations (1977). That is why he introduces the factor of time, denoted by t, and then distinguishes two types of motion of an object category: *unfolding* and *transformation*. The former is defined as follows. Given the object category at time t

$$K_E^t = (U^t, S_E^t, N_E^t) \tag{4.4}$$

where the structure at this time is as follows

$$
\begin{array}{ll}
S_E^t & G \\
 & G, p_1 \\
 & G, p_1, p_2
\end{array}
$$

and the nomological structure N_E is given through f, g, and h, so that

$$
\begin{aligned}
E^t &= f(G) \\
E^t &= g(G, p_1) \\
E^t &= h(G, p_1, p_2)
\end{aligned}
$$

Let us assume such a change in an object that at time $t+1$ it holds

$$
\begin{array}{ll}
S_E^{t+1} & G \\
 & G, r_1 \\
 & G, r_1, r_2
\end{array}
$$

with the nomological structure N_E given through f, k, l, so that

$$
\begin{aligned}
E^{t+1} &= f(G) \\
E^{t+1} &= k(G, r_1) \\
E^{t+1} &= l(G, r_1, r_2)
\end{aligned}
$$

S_E^{t+1} is in this type of a change in "principle similar to, yet not identical" (Nowak 1977, 111) with the structure S_E^t, and the nomological structure N_E^{t+1} "proceeds from the same regularity as the nomological structure" N_E^t "though, beginning from a certain point they differ in the form of manifesting this regularity. Put otherwise: the same regularity is manifest in other ways . . . the new nomological structure differs from the previous one only in additional forms of manifestation" (Nowak 1977, 112–113).

Transformation is the sort of motion of the object category where "the hitherto repertoire of the principal factors undergoes changes, hence new regularities start to determinate the examined phenomena" (Nowak 1977, 125). Let me suppose that in time t the structure (4.2) and the nomological structure (4.3) hold, and that at time $t+1$ it holds that:

$$
\begin{array}{ll}
S_E^{t+1} & H \\
 & H, r_1 \\
 & H, r_1, r_2
\end{array}
$$

with the nomological structure as follows

$$
\begin{aligned}
E^{t+1} &= k(H) \\
E^{t+1} &= l(H, r_1) \\
E^{t+1} &= m(H, r_1, r_2)
\end{aligned}
$$

Thus there occurs the transformation of the object category expressed by (4.4) into

$$K_E^{t+1} = (V^{t+1}, S_E^{t+1}, N_E^{t+1}) \tag{4.5}$$

Naturally, a fundamental question immediately arises. Why does the principal factor (essence) G transform into a new principal factor (essence) H? Nowak's answer can be found in his attempt to join the concept of unfolding with that of transformation. This unification is represented by him as follows (1977, 190):

$$K_E^0 \, UN \, K_E^1 UN \ldots UN \, K_E^j \, TR \, K_E^{j+1} \tag{4.6}$$

The transformation, *TR*, should occur as a result of the inceasingly recurring unfoldings, *UN*, that is, as a result of recurring changes of the modification conditions (see also Nowak and Nowak 1985, 265).

It is noteworthy that Nowak addresses the transformation of an object category in connection with Marx's transition from the law of value to the law of surplus-value in the first volume of *Capital*. The structure of value, Nowak believes, can be represented as follows:

$$V_G$$
$$V_G, F_G$$

where V_G denotes the value of gold and F_G fluctuations of the demand and supply of gold. The golden money, he argues, converts into capital "when it buys labor-power which produces surplus-value" (1977, 127). The structure is (m denotes surplus-value)

$$V_G, m$$
$$V_G, m, F_G$$

and the change of V_G into V_G, m, as a change of a principal factor, represents a transformation of the object category. Here it is evident that no quantitative changes of F_G can bring about the transition from V_G to V_G, m because the transformation takes place at an ontical level of principal factors, i.e., where the modification condition, denoted as F_G, is not functioning at all. The same holds good for both the prediction and retrodiction in Part VIII of the first volume of *Capital*. In the case of retrodiction, Marx's progression of thought may be expressed as the transition from K_E^t towards K_E^{t-j} ($j \geq 1$); in the case of prediction as the transition from K_E^t towards K_E^{t+j} ($j \geq 1$). Both progressions, however, have been done without reference to the modification conditions. Marx's thought operations of prediction and retrodiction here were not based on the description of the cumulative changes of secondary factors, since he eliminated them completely from all thought operations involved in the first volume of *Capital*. They are here neither zero nor non-zero, because the respective forms of manifestations (e.g., profit) they modify have yet to be derived.

Nowak also defines the concept of (natural) kind and of sub-kind, where the former "is such an aggregate of all objects for which is given the same principal factor (or principal factors)" (1977, 81), and the latter is such an aggregate of all objects with the same essence and the same modification condition.

By utilizing the ontological concepts introduced by Nowak it is possible to provide an ontological foundation of scientific laws of the $L^{(k)}$ type. Since $f_1(H) = E^{(k)}$ holds here, one can represent the ontical levels (the internal level, . . . , the surface level) as follows:

$$
\begin{array}{ll}
(k) & f_1(H) = E^{(k)} \\
(k-1) & E^{(k-1)} = E^{(k)} + g_k(C_{mod_k}) \\
\vdots & \vdots \\
(0) & E^{(0)} = E^{(k)} + g_k(Cmod_k) + \ldots + g_1(Cmod_1)
\end{array}
$$

Is such an ontological foundation complete? A view of an object (its ontology) is complete if, and only if, at each level of scientific explanation, the respective form of manifestation is *necessary* for this object given the impact of the respective modification conditions. So, in the case of Popper's moa example this means that

All moas must necessarily die at the age of fifty years given the impact upon them of viruses of such-and-such type.

But this presupposes that at a certain level, let me denote it as k', it is already intelligible why the ground-essence-cause H manifests itself necessarily as the effect-grounded $E^{(k)}$. In the case of the moas it would mean that it is known why their biological structure, in ideal (favorable) conditions, manifests itself having a life-expectancy of at least sixty years. But this in turn presupposes knowledge at an even deeper level, let me denote it as k'', of the very essence H, independently of all phenomena. But, as I have already shown, knowledge based on scientific laws of the type $L^{(k)}$ does not contain any such type of knowledge about the level k'' and therefore no knowledge about the level k'. So the view of objects (the ontology) provided by scientific laws of the type $L^{(k)}$ is in principle incomplete.

The incompleteness of this ontology also becomes evident when one scrutinizes the relation of the property denoted by the predicate letter N and of the property denoted by the predicate letter H through two questions. First, what is the relation between the properties denoted by N and H in one entity? In subchapter 3.4 I claimed that a scientific law of the type $L^{(k)}$ should be non-vacuously (non-trivially) fulfilled by an entity of a certain kind which we have reduced in mind to the essence of this kind. Therefore, the answer should be as follows. Each entity having H is necessarily from N. But when one poses this question to (3.9), expressing $L^{(k)}$, this formula is reduced to

$$(x)(Nx \rightarrow Hx) \tag{4.7}$$

This formula does not express the relation of the properties denoted by N and H as necessary one. It does not exclude the case that there are possible worlds inhabited by entities not from N but still having the (essential) property H.

The second question can be posed as follows. Why do all instances (entities) of N have the same essence H? The ontology upon which $L^{(k)}$ rests does not provide any answer. It just presupposes, without any foundation, that all entities belonging to the same kind also have the same essence.

The conclusions of chapter four are as follows. First, the incompleteness of the ontology reconstructed above indicates the need for a new ontological foundation consisting of a more developed type of scientific law as compared to that of $L^{(k)}$. Second, the existence of questions that cannot be answered in the framework of that ontology makes it clear that its conceptual "outfit" provides no means for the answers to be found. Those questions are posed correctly, but they are posed in an inappropriate framework; they do not match each other. They overstrain the conceptual capabilities of this framework and should therefore be posed and answered in a more developed conceptual framework. Its creation will be the task of part three.

PART 3

THE INHERENT LAW OF ESSENCE

. . . inherent organic laws of political economy now at work in every civilized town (Marx 1984, 252; Marx and Engels 1975ff., Vol. 12, 222).

The capital is the principal sum, the profit is the incidental sum (Nebensumme) produced by this principal sum in a certain time of circulation. The principal sum, the capital, bears a proportion as the ground (cause) to that incidental sum as the grounded (consequence, effect) (Marx 1980d, 1625; Marx and Engels 1975ff., Vol. 33, 96).

The progress of our investigation will show that exchange-value is the only form in which the value of commodities can manifest itself or be expressed. For the present, however, we have to consider the nature of value independently of this, its form. . . . In fact we started from exchange value, or the exchange relation of commodities, in order to get at the value that lies behind it. We must now return to this form under which value first appeared to us (Marx 1990, 32, 40; 1967, Vol. 1, 38, 47; Marx and Engels 1975ff., Vol. 35, 48, 57).

CHAPTER 5

THE INHERENT LAW OF ESSENCE

Hegel in the *Phenomenology of Mind* views the pure law as a genuine scientific law and regards it as a preparation for a higher level of scientific knowledge, where attempts at formulations of a more developed law are made. Hegel labels it "the organic law" and discriminates between its three forms: a scientific law expressing the relation among the organic and the non-organic (the elements); a scientific law expressing the relation of the inward and outward (form) of the organic; and, finally, a scientific law expressing the internal characteristics of the organic (irritability, sensibility, reproduction). Gradually, however, he jettisons all these three varieties of an organic scientific law and arrives at the conclusion that "the representation of a law is lost altogether in the case of the organic" (1928, 207; 1967, 310). Marx, however, is of the opposite opinion on this point. Already in an 1853 article he mentions the "inherent organic laws of political economy now at work in every civilized town" (1984, 252; Marx and Engels 1975ff., Vol. 12, 222).

5.1 THE INHERENT CONDITIONS, THE INHERENT LAW AND THE INHERENT IDEALIZATIONAL TYPE OF A SCIENTIFIC LAW

Marx reconstructs conditions different from modification conditions and labels them explicitly "essential conditions." He uses this term with respect to the production of surplus-value, *SV*, from surplus-labor, *SL*, and lists them as follows (1981a, 371–372; Marx and Engels 1975ff., Vol. 28, 391):

(1) On the one side, the existence of living labor capacity as purely *subjective* existence, separated from the moments of its objective reality; therefore separated just as much from the *conditions* of living labor as from the *means of existence*, the *means of subsistence*, the means of self-maintenance of living *labor-capacity*; the living possibility of labor on one side in this complete abstraction. (2) On the other side, the value or objectified labor must be an accumulation of use values, sufficiently large to provide the objective conditions not merely for the production of products or values necessary to reproduce or maintain living labor capacity, but also to absorb surplus-labor, to supply the objective material for the latter. (3) A system of free exchange – money circulation – between the two sides; a relation between the two extremes which is based upon exchange values, not on the lord-subject relationship, i.e. production which does not directly supply the means of subsistence to the producers but is mediated by exchange, and which cannot therefore usurp alien labor directly, but must buy it, exchange it, from the worker himself. Finally (4) the one side – which represents the objective conditions of labor in the form of independent values-for-themselves – must present itself as *value*, and must regard as its ultimate aim the positing of value, self-valorization, the creation of money – not immediate enjoyment or creation of use value.

That it is not a case of randomly chosen conditions is readily seen from the fact that points (1) to (4) were included without alteration in the second manuscript of *Capital* (*Manuscripts 1861/63*), where Marx referred to them again explicitly as essential conditions (1982a, 2287; Marx and Engels 1975ff., Vol. 34, 246). Let me label these conditions, following Černík's (1977), as "immanent (essential, inherent) conditions" and denote them as follows

Cim_1x – x is a free owner of his own capability to work, separated from the means of production and subsistence;
Cim_2y – y is the owner of the means of production and subsistence;
Cim_3x,y – the relation of x and y is that of buyer and seller, of free exchange;
Cim_4y,x – the anticipated aim of y with respect to x is a self-increasing value.

The law of surplus-value, as given in the first volume of *Capital* prior to the introduction of the conceptual differentiation between constant and variable capital, has the form

$$L_{SV} : (x)(y)(Cim_1x \,\&\, Cim_2\,y \,\&\, Cim_3x, y \,\&\, Cim_4y, x \underset{T}{\rightarrow} SVx, y = SLx, y) \qquad (5.1)$$

SVx,y denotes the surplus-value created by x for y and SLx,y denotes the surplus-labor done by x for y, and where the latter is the *substance* of the former (Marx 1992, 221; 1967, Vol. 3, 149). Formula (5.1) states that if all the four essential conditions are given, then surplus-value is necessarily produced, so that it equals the expended surplus-labor.

The concept "essential (immanent) conditions" is also a key to the reconstruction of the scientific law of value in the first volume of *Capital* as it stands prior to the introduction of the concepts of exchange-value, price, etc. The immanent conditions, Cim_0, for the production of value from socially necessary labor is the existence of separated producers who own the means of production. The law of value can be written as

$$L_V : (x)(\,y)(Cim_0x \underset{T}{\rightarrow} Vy = g(Lx, y)) \qquad (5.2)$$

If x satisfies the condition Cim_0, then x's labor, L, necessarily leads to the production of the product y with value V, i.e. to the creation of a commodity.

In (5.1) and (5.2) I use the symbol "$\underset{T}{\rightarrow}$". Its introduction represents yet another departure from Nowak's reconstruction of Marx's law of value from the first volume of *Capital*. L. Nowak uses in (3.1) the symbol "$\longrightarrow$" for the sentential connective "If . . ., then . . .", and which should appear in the the language stating the relation between price and value as it is given in the first volume of *Capital*. In my view, Nowak's approach employing that sentential connective cannot be used in the reconstruction of Marx's understanding of the relation of price to value or in the reconstruction of his understanding of the relation of value to socially necessary labor because the language the latter uses to express both contains *alethic modalities*. That this is so can be readily seen from the following two statements of Marx: "[t]he magnitude of value of a commodity expresses . . . a necessary, to its process of production inherent relation to the social labour time" (1987b, 129); "Money as a measure of value is the necessary form of manifestation of the immanent measure of value of commodities, the labour time" (Marx 1987b, 121). What I do not develop in this book is a syntactical and semantical analysis of the sentential connective symbolized by "$\underset{T}{\rightarrow}$".

I understand the scientific law of value and the scientific law of surplus-value from the first volume of *Capital* as exemplifications of the inherent (immanent) type of scientific law symbolized as:

$$L_i : (x)(Cim_{1-l}x \underset{T}{\rightarrow} Hx = f_2(Sx)) \quad (5.3)$$

Cim_{1-l} denotes a conjunction of 1 to l ($l \geq 1$) essential conditions under which the essence-producing process is necessarily triggered, i.e., under these conditions the essence H necessarily arises out of a certain essence-producing substance S. A scientific law of the type L_i expresses the production of the essence of instances of a natural kind, and therefore of a certain natural kind as a whole. It expresses only the necessary characteristics of the kind and of its instances.[1] The non-necessary, accidental characteristics, which may vary from sub-kind to sub-kind, have yet to be derived; a law of the L_i type is only a law of the "pure" kind.

Knowledge progresses from the scientific inherent law[2] towards a scientific law of the kind where the consequent already includes an equation with the structure $E^{(k)} = f_{3k}(H)$, i.e., the phenomenal form $E^{(k)}$ is already defined through essence, and, therefore, already has the character of the *manifestation of essence*. This is the kind of scientific law I shall label the inherent idealizational type of scientific law, and it has the character of the inherent law of the manifestation of essence. In symbols

$$L_{ii}^{(k)} : (x)(Cim_{1-l}x \,\&\, Cmod_{1-k}x = d_{1-k} \underset{T}{\rightarrow} E^{(k)}x = f_{3k}(Hx)) \quad (5.4)$$

While taking into account the definition of the essence in (5.3) it holds

$$L_{ii}^{(k)} : (x)(Cim_{1-l}x \,\&\, Cmod_{1-k}x = d_{1-k} \underset{T}{\rightarrow} E^{(k)}x = g_k(Sx)) \quad (5.4')$$

where g_k denotes a compound function so that holds $g_k(S) = f_{3k}(f_2(S))$. $L_{ii}^{(k)}$ is exemplified by the law of value at the beginning of the manuscript of the third volume of *Capital*.

5.2 THE INHERENT ORGANIC, THE DEVELOPMENTAL, AND THE HISTORICAL TYPES OF SCIENTIFIC LAWS

Marx does not confine himself to the understanding of the law of surplus-value as a scientific inherent law of essence. He also grasps it as a law regulating the process of reproduction. In his view "[w]hatever the form of the process of production in a society, it must be a continuous process, must continue to go periodically through the same phases . . . When viewed, therefore, as a connected whole, and as flowing on with incessant renewal, every social process of production is at the same time a process of reproduction. The conditions of production are also those of reproduction" (1990, 493; 1967, Vol. 1, 566; Marx and Engels 1975ff., Vol. 35, 565). The law of surplus-value in this form exemplifies the scientific inherent organic type of law. The latter describes not only the initiation of the essence-producing process under definite essential conditions, but also how these essential conditions, through the very essence-producing process, are renewed-reproduced, a renewal which in turns brings about further initiations of the essence-producing process. Reproduction is here understood as a simple process, i.e., there is no change in the quantitative determinations of the essential conditions nor in those of the essence. In symbolic form it is expressed as follows:

$$L_{io} : (x)(Cim_{1-l}x \underset{T}{\twoheadrightarrow} Hx = f_2(Sx) \underset{T}{\twoheadrightarrow} Cim_{1-l}x \underset{T}{\twoheadrightarrow} Hx = f_2(Sx) \ldots) \qquad (5.5)$$

A scientific law of the type L_{io} enables us to grasp the functioning – duration through self-preservation – self-motion of objects with the essence H. Since scientific laws of the L_{io} type express the process of reproduction, they fix a certain type of regularity. This type of a scientific law can therefore be viewed as a reinterpretation of scientific laws formulated according to the Humean regularity approach. But here regularity is not understood as the recurrence of the succession of (empirically observable) events, but as reproduction at the level of the very essence, at the level of the structure underlying the observable events-phenomena, and where this structure and the reason of its persistence, namely, its self-reproduction, are comprehensible only theoretically.

Formula (5.5) represents a description of the functioning and duration of a "pure" natural kind, i.e., the latter's self-preservation is not as yet grasped through a description of the mutual interactions of its subkinds.[3] What's more, this formula can serve as a basis for the comprehension of the so-called essential formality. Marx introduced this concept in (1988b, 129; Marx and Engels 1975ff., Vol. 34, 465) to explain why the production process of surplus-value is constantly renewed. For Marx the essence-producing process is connected in a cyclical way with the process of essential formality (the circulation process). Together they set up an operationally closed system, and the reproduction of the essential conditions results from the reciprocal alternate recurrence of these two processes.

The above represents an ontic state of affairs. Beginning with an epistemic approach, Marx starts from the description of the essence-producing process; in fact he proceeds from the scientific inherent law (the description of capital's conversion into surplus-value). He then turns to the essence-producing process as a process of reproduction (the realization of the commodity containing objectified surplus-value and the recurrent initiation of the production process); this, however, makes the description of the process of essential formality indispensable. That is why Marx assumes that the latter is given between two subsequent essence-producing processes, operating "in its normal way" (1990, 492; 1967, Vol. 1, 564; Marx and Engels 1975ff., Vol. 35, 564). Hence, he concludes, a description of the process of essential formality is not necessary in the account of the very essence-producing process (Marx 1967, Vol. 2, 23; Marx and Engels 1975ff., Vol. 36, 31). But when the description of the essence-producing process as a reproduction process has been finally accomplished, it then serves as an explanatory foundation for the description of the very process of essential formality (the transition from the description of the production process of capital in the first volume of *Capital* to the description of the circulation process in the manuscript of its second volume).

Marx does not stop at the description of simple reproduction, but moves on towards the description of enlarged reproduction (change of surplus-value back into capital). I see the scientific law of surplus-value in this form as an exemplification of the developmental type of a scientific law. It describes the reproduction process as that of non-simple reproduction, i.e., grasps the quantitative changes of the essence, of the essence-producing substance, and of the essential conditions which take place as the cycles of reproduction succeed one another. It can be put down as follows:

$$L_d : (x)(Cim_{1-l}x \underset{r}{\rightarrow} Hx = f_2(Sx) \underset{r}{\rightarrow} Cim'_{1-l}x \underset{r}{\rightarrow} H'x = f_2(S'x) \underset{r}{\rightarrow} Cim''_{1-l}x \underset{r}{\rightarrow} \ldots) \quad (5.6)$$

A scientific law of the type L_d enables not only the description of the functioning-duration-self-motion of objects with the essence H, but also their self-development; it is no longer symmetrical to time reversal.

In the first volume of *Capital* the law of surplus-value also has a form that qualifies it as a case of the scientific historical law. It is this type of law which involves the description of the historical process which brought about the rise of the essential conditions (symbolized here as ${}_hCim_{1-l}$) for the existence of the essence H. It also includes the description of the points of non-simple reproduction where the process of the decline of essential conditions is taking place, symbolized here as $Cim_{1-l} \Rightarrow 0$, and, therefore, where the essence is fading away. The schema is as follows:

$$L_h : (x)({}_hCim_{1-l}x \underset{r}{\rightarrow} Hx = f_2(Sx) \underset{r}{\rightarrow} Cim'_{1-l}x \underset{r}{\rightarrow} H'x = f_2(S'x) \underset{r}{\rightarrow} \ldots \underset{r}{\rightarrow} Cim_{1-l}x \Rightarrow 0 \quad (5.7)$$

Here ${}_hCim_{1-l}$ denotes that 1 to l essential conditions have come to exist historically; their becoming is here as yet only a result of the self-development of a prior natural kind. The new essence of the new natural kind originates from these conditions, and this new natural kind by its own movement, on its own ground, changes, reproduces ${}_hCim_{1-l}$ into its own result. Marx had such a change in mind when he claimed that "[t]he conditions and suppositions of the becoming, origination of capital impute precisely that it is not as yet, but only becomes; hence they disappear with the real capital, capital which, starting from its own reality, sets the conditions of its own realization" (1981a, 368; Marx and Engels 1975ff., Vol. 28, 387–388). The essential conditions of the production of surplus-value, denoted by me as Cim_{1-4}, with respect to the capital, "appear now as the result of its own realization, reality, set by it – not as the conditions of its origination, but as the result of its existence" (1981a, 368; Marx and Engels 1975ff., Vol. 28, 388).

A scientific law of the type L_h is not only a description of the functioning/duration/self-motion/self-development of objects with the essence H, but also a description of their *origination* (out of other objects with a different essence) and of their *decline* (transformation into other objects with a different essence). L_h therefore describes the rise and decline of a "pure" natural kind. But this in turn means that it is possible to pick out "the points where a historical analysis ought to be applied," which would, on the one hand, permit us to reconstruct, to retrodict "the past lying behind this system" and, also, to predict the points "foreshadowing the future, of the becoming motion" (Marx 1981a, 369; Marx and Engels 1975ff., Vol. 28, 389). A law of the L_h type provides a basis for both the prediction and retrodiction procedures which lead scientific knowledge beyond the borders of a given natural kind because it facilitates comprehension of the natural kind from which it really arose in the past and comprehension of the natural kind into which it will transform itself in the future.

The reconstruction of the L_h type of scientific law sheds new light on the problem of the evolution of laws of nature treated earlier in this century by H. Poincaré (1911) and more recently by Y. Balashov (1992). The latter suggests two "ontological" examples, drawn from natural science, of a possible evolution of real laws in objective reality. Chemical laws hold between certain objects capable of chemical interaction, so (Balashov 1992, 346)

> the laws of chemical phenomena could not come into existence until the objects had been formed that were capable of interacting chemically. It happened obviously during the definite cosmic epoch when the temperature fell sufficiently to enable formation and stable existence of neutral atoms and then molecules of various substances. Prior to this epoch there were no chemical laws. Hence it is reasonable to speak of their coming into being at some stage in cosmic evolution.

Similarly, one can suggest an "evolution of evolution" in the world of living objects (Balashov 1992, 346–347)

> It is quite natural to suggest that in the process of progressive biological evolution its factors and moving forces (i.e., heredity, population dynamics, reproductive ability, trophic relations, etc.) are subject to a change no less than its products. The evolution of factors (as well as products) proceeds under the pressure of selection.

These examples lead Balashov to the central issue. How should scientists reproduce in scientific laws the evolution of laws of nature? Balashov proposes the following principle (1992, 355)

> The constancy of laws in the description of nature may always be saved by the transition to a deeper level of fundamentality. Though no fundamental law of nature is insured against possible mutability, actual discovery of the latter cannot be 'ratified' unless a more general law of science is 'ready' that regulates the evolution of the former law while being itself immutable. The former law loses thereby its fundamental status.

So when scientists find out that there exists a space-time region where a certain law does not hold (e.g. chemical laws in the very early universe), they view the objects for which these laws were formulated as belonging to level n, which is only a manifestation of a "deeper" level $n + 1$, a subordinate case of it. The basic idea implied by the above-mentioned postulate is an incessant *cognitive move* from scientific laws formulated for certain kinds of entities to a scientific law formulated for a 'higher-level' kind of entities, with respect to which the previous kinds appear only as its subkinds. Balashov, therefore, correctly relates this "layer-upon-layer" model of reality with "its natural counterpart, the universal 'covering law' explanatory structure" (1992, 364). This is consistent with the idea that certain conditions should enable scientists to explain all properties known previously about level n on the basis of the law for level $n + 1$. So as he operates only with the concept of initial conditions (1992, 347, 354), Balashov's argument comes to resemble arguments based on the D-N model of scientific explanation. Surprisingly, in the more general ontological reflections on the evolution of laws of nature, he claims something very different from the D-N model, namely, that "the laws of nature, that is, the systems of essential relations, are embedded in the 'world of essences' that is subject to development as well as the 'world of phenomena"' (1992, 347).

The idea of a theoretical unification of structurally different objects of reality based on a scientific historical law sacrifices the usual "tool of unification" – the idea of a relation of subkinds to their kind. Instead it bases its attempt to discover and explain unification on the idea of development. According to this proposal, the aim of scientists calling on their knowledge in order to pursue a unification attempt should not be to find exclusively a more profound structural level $n + 1$, of a more universal law, which instantiates level n and its laws, but also to find unification in a *developmental manner*, to explain theoretically how one *kind* through its self-development transforms itself into another kind with a different essence. The driving idea is to give up the search for eternal determinations reappearing on different kinds of objects, which are then fixed in a more profound scientific law, and instead to formulate scientific historical laws which hold only for these kinds and to mediate between them on the basis of these laws.

The scientific law of surplus-value from the first volume of *Capital* initially has the form of the scientific inherent law (the description of the rise of surplus-value together with the description of all the aspects of this rise in Part III to Part VI). Later, it has the form of the scientific inherent organic law (the description of the recurrent initiation of the production-process in its quantitatively immutable scope, i.e., the description of its simple production in Chapter XXIII), then the form of the scientific developmental law (the theoretical reconstruction of the cycle capital → surplus-value → capital, i.e., the description of the enlarged reproduction and the issues related to it in Chapters XXIV and XXV); and finally the form of the scientific historical law. It is on this basis that Marx performed both the prediction and retrodiction in Chapters XXVI to XXXII from Part VIII of the first volume of *Capital*.

When one takes into account our discriminations between these types of law – pure idealizational law, the inherent law, the inherent idealizational law, the inherent organic law, the developmental law, and the historical law – it becomes clear that one cannot accurately designate Marx as the "Galileo of social science" on the basis of his economical writings, where he allegedly formulated scientific (pure) idealizational laws, for in reality he formulated richer, more advanced types of scientific laws than the pure idealizational type of scientific law.

Let me now examine the problems posed by Marx's own passage from the scientific law of value to the scientific law of surplus-value as it is given in the first volume of *Capital*. In (5.2) I have formulated the logical structure of the scientific law of value, which expresses the rise of the instances of the natural kind of commodities. At this point Marx asserts that "[w]hat I start from is the simplest social form in which the labor product is represented in contemporary society" (1969, Bd. 19, 369; Marx and Engels 1975ff., Vol. 24, 544), i.e., "we proceed from it as the most simplest element in capitalist production" (1988b, 30; Marx and Engels 1975ff., Vol. 34, 362). Thus in thought Marx moves in the framework of one natural kind, whose essence still has to be determined. The scientific law of value constitutes a foundation for the production-relations between the producers-owners of commodities, and for the exchange of commodities for prices equal to their values (in the case that the demand equals the supply and there is no monopolistic domination of technologies, raw materials, etc.). The existence of the producers-owners of a commodity has been assumed by the explicit statement of the condition Cim_0 in the formulation of the scientific law of value.

One can arrive at the formulation of the scientific law of surplus-value by presupposing the existence of such a commodity that possesses a specific property, namely its potential to create in the process of its productive consumption a value exceeding its own. This is labor power (capability to work) considered as a commodity. To proceed from the scientific law of value to that of surplus-value it is necessary to *include* in the explanation a description of the existential conditions of labor power as a commodity, i.e., conditions I have labeled $Cim_{1,2,4}$; the condition Cim_3 being already involved in the formulation of the law of value as Cim_0. In symbols the explanatory move looks as follows (with "$=_{tr}= |$" standing for transformation)

$$L_V \,\&\, Cim_{1,2,4} =_{tr} = |L_{SV} \qquad (5.8)$$

The above accounts for the fact that the scientific law of value, taken already as a law of the type $L_{ii}^{(k)}$ (i.e. with the equation relating price to value) has three idealizations in its early state, when it is posed prior to the accomplishment of the passage to the scientific law of surplus-value in the first volume of *Capital*, whereas in its later state, at the beginning of the manuscript of its third volume, there are more than three idealizations. The reason for this lies in the fact that the *concept of surplus-value and the explanation of all its metamorphosis have come to be reckoned with.* Their relation to the structure of the law of value in the manuscript of the third volume of *Capital* is as follows. After having derived in the first volume the concept of surplus-value, *m*, and of the rate of surplus-value, *m'*, Marx turns in the manuscript of the third volume to describe its metamorphoses in the following order. In "subchapter" one of the first chapter "The Conversion of Surplus-Value into Profit" (1992, 7–50), he first nominally sets the period of one year as a standard (*Maßstab*) for one turnover of capital and then moves from *m* and *m'* to the concept of profit and to that of rate of profit. In "subchapters" one, two, and three from the second chapter "Conversion of Profit into Average Profit" (1992, 212–273), he inquires about the results caused by the difference of the average organic composition of capital in separate branches of industry and explains the process of the emergence (as a result of competition) of the average profit, *p'*, and of the average rate of profit, *p''*. All these reasonings have been made by Marx on condition that there exists only industrial capital (in the form of money-, commodity-, and production-capital).

The situation changes in the first "subchapter" of the fourth chapter "Conversion of Commodity-Capital and Money-Capital into Commodity-Dealing and Money-Dealing Capital or into Merchant-Capital," where Marx describes the autonomization of the commodity-capital, i.e., that the selling of the commodity (the latter being the vehicle for the average profit) becomes the job of an independent social group of commodity merchant capitalists who appropriate the merchant profit m_p. In such a way Marx arrives at the explanation of the *innate* splitting, differentiation of the average profit. This explanation does not match up with Nowak's characterization of this process as one imposed *a priori*, from the outside. Average profit is now given as the sum of the industrial profit, i_p, and the merchant profit, i.e., $p' = i_p + m_p$. A further step consists in the consideration (Chapter 5: "Splitting of Profit into Interest and Enterprise-Profit") that neither the merchant nor the industrial capitalist operates with his own capital. The existence of a social group that loans money-capital with the intention of getting it back with a profit – *interest* – is now being assumed. Thus p' is again innately differentiated, now into enterprise profit, e_p, which represents a quantitative modification of the sum $i_p + m_p$, and the interest, i, so that holds $p' = e_p + i$. And, lastly, Marx reflects upon average profit under the conditions of capitalist agriculture (Chapter 6: "Transformation of Surplus-Value into Ground-Rent").

On the basis of his explanation of all these quantitative modifications of surplus-value, Marx also undertakes quantitative modifications of the law of value. He shows how the quantitative changes of m bring about the quantitative determination of both value and price of a capitalistically produced commodity. Considering the distinction between the organic composition of capital in different branches of production and within each branch, he derives the concept of production price and market-value ("subchapters" 1, 2, and 3 of the second chapter). Proceeding from this point, he explains the sale of a commodity as enabling separate groups of capitalists to appropriate their respective profits (first and second "subchapter" of the fourth chapter) and finally makes an account of the problems involved in the price-formation of agricultural products (sixth chapter).

The above considerations clarify the prima facie surprising fact of the divergence between the number of idealizations required for the scientific law of value at the beginning of the first volume of *Capital* and at the beginning of its third volume. By means of the concept of surplus-value one arrives at the condition denoted by Nowak as p_3. By means of the concept of surplus-value one arrives at the concept of capital and therefore also at the concept of the organic composition of capital (i.e., at the conditions denoted by Nowak as p_4, p_6, and p_8). Through surplus-value, by means of the explanation of its split, it is also possible to arrive at the condition denoted by Nowak as p_5.

5.3 TWO TYPES OF SCIENTIFIC EXPLANATION BY GRADUAL CONCRETIZATION

Having distinguished between the pure idealizational type of scientific law of essence, $L^{(k)}$ – expressed by (3.9) – and the inherent idealizational type of scientific law of (the manifestation of) essence, $L^{(k)}_{ii}$, expressed by (5.4), one now has to ask the following important question. Are the whole course and aims of scientific explanation proceeding from a scientific law of the $L^{(k)}$ type identical to the course and aims of scientific explanation proceeding from a scientific law of the $L^{(k)}_{ii}$ type? To answer this question it is necessary first to determine the origin of the secondary (modification) conditions as they have been reconstructed by Nowak in (1980). The condition $p_1 = 0$ implies that there exists a free motion of all the components of the production process (labor power, production technologies, natural resources, etc.) between branches of production as well as within each of them, i.e., that no enterprise can forever monopolize this or that production component. Any component that would be accessible only to one or few producers (e.g. a spring of hot water covering the energy input) would represent a *value* (price) advantage to those producers against the remaining producers who are deprived of access to this component. So, in my view, condition p_1 is a function of value, and therefore also of socially necessary labor. Generally speaking, one can claim that this condition describes certain processes going on between subkinds within one natural kind and is a function of the essence-producing substance. The same applies to the condition p_2 expressing the interrelation, through a circulation of commodities with a certain *value*, between a national economy and other economies. Condition p_3 describes the quantitative distinction of the rate of surplus-value generated by a certain enterprise with respect to the average rate of surplus-value in the national economy as a whole. So, again, one deals here with a value-dependent (therefore also with an essence-producing substance-dependent) condition expressing the quantitative determinants of a subkind which distinguishes the latter from the quantitative determinations of the whole kind. The same holds also for conditions p_4, p_6, and p_8, which express a quantitative distinction of the organic composition of capital in agriculture as against the average organic composition of capital in the rest of economy, or, again, quantitative distinctions between one enterprise compared with the average organic composition of capital in the branch to which it belongs, or, finally, quantitative distinctions between one branch of the economy as it compares to the whole of the national economy. Because the organic composition of capital represents its value-composition, and value is a function of the socially necessary labor, the conditions p_4, p_6, and p_8 are functions of the substance of value. Similarly, the condition p_5, expressing the merchant profit, is a function of the surplus-labor because the former is only a fraction of surplus-value. Working from these statements, one concludes that the conditions p_1 through p_6 as well as condition p_8 are generated by the quantitative determinations of the subkinds and are functions of the essence-producing substance S. So for the modification conditions it holds (l_k denotes a function, $k \geq 1$)

$$Cmod_k = l_k(S) \tag{5.9}$$

A more intricate functional dependence is given for the condition p_7, expressing the difference between the supply and the demand. On the one hand, if division of total social labor in the national economy as a whole is given, and the wants and incomes of the population are given, and the productivity of labor is given, then a given branch of the economy produces commodities in a certain amount of a certain value. Supply is here a function of value and hence also a function of its substance. On the other hand, the amount of the demand, given the same division of total social labor, the same wants and incomes, and the same productivity of labor, is again a function of the value of a given amount of commodities, and therefore also of its substance as well.

But, simultaneously, the demand is dependent on human wants, where these wants depend on non-economic and non-social factors. Therefore, these wants may be altered by non-economic and non-social changes. So, for example, a lasting and rapid decline of the earth's temperature would generate a profound change in consumption patterns. The demand for energy would increase without any change in the labor-productivity in energetics or in any other branch of production. A modification condition, which is not only a function of the essence-producing substance S but also of parameters independent from the latter, will be symbolized as follows ($m \geq 1$, $r \geq 1$):

$$Cmod_r = l_r(S, e_1 \dots, e_m) \tag{5.10}$$

where e_i denotes an external, with respect to S, parameter (on such a type of condition see (Mandel 1985)).

If we speak of the advance of scientific knowledge, then this suggests that the step succeeding the formulation of the scientific law of the L_i type is not only the formulation of a scientific law of the $L_{ii}^{(k)}$ type grasping the genesis of the manifestation of essence $E^{(k)}$, but also the description of the genesis of the modification conditions together with the comprehension of their functional dependence on the essence-producing substance and external parameters.

The fact that $E^{(k)}$ is expressed in (5.4′) as a function of S has, together with the knowledge of the functional dependence of the modification conditions, a decisive impact on the unfolding and the result of explanatory procedures beginning with a scientific law of the type $L_{ii}^{(k)}$. If the equation $E^{(k)} = g_k(S)$ holds, then from the character of the explanatory procedure by gradual concretization it follows that all forms obtained from $E^{(k)}$, by taking into account the impact of the modification conditions of the type (5.9) are functions of S. That is, since

$$E^{(k-j)} = f_{k-j}(H(S), Cmod_k(S), \dots, Cmod_{k-(j-1)}(S)) \tag{5.11}$$

holds, then

$$E^{(k-j)} = g_{k-j}(S) \tag{5.11′}$$

also holds. So, for example, by abolishing the idealization $p_8 = 0$, value is transformed into market-value, where the latter is again a function of the socially indispensable labor. When abolishing $p_6 = 0$, surplus-value changes into average profit, which is also a function of surplus-labor. Simultaneously, value changes into production price, which is a function of the same substance as value.

A different situation obtains when conditions of the type (5.10) are used in an explanatory procedure by gradual concretization. The derived forms of manifestation $E^{(k-1)}, \dots, E^{(0)}$ are then not only a function of S but also of the external parameters. That is, since

$$E^{(k-j)} = f_{k-j}(H(S), Cmod_k(S, e_m), \dots, Cmod_{k-(j-1)}(S, e_1)) \tag{5.12}$$

then

$$E^{(k-j)} = h_{k-j}(S, e_m, \dots, e_1) \tag{5.12′}$$

reproduction of commodities in every industrial sphere of production" (Marx 1992, 272; Marx 1967, Vol. 3, 198). It is not surprising that Marx interprets the law of the production price as expressing a reproduction relation. In his approach (1988a, 15; Marx and Engels 1975ff., Vol. 34, 345)

[e]very *price* must be reducible to a *value*, because the price is in and for itself only the monetary expression of value. And the circumstance that actual prices stand over or below the price corresponding to their value, does not make any difference for the fact that they are a – even if in this supposed case a quantitatively too big or to small – *quantitatively* incongruent expression of the value of the commodity.

and (1992, 426; 1967, Vol. 3, 354)

[p]rice, after all, is the value of commodity (this is also true of the *market-price*, whose difference from value is not one of *quality*, but only one of *quantity*, referring only to the quantity of value (Werthgrösse) . . . A price which differs from value in quality is a contradictio in terminis.

So the same holds for the production price. It is also only a *quantitative* modification of the magnitude of value, and so, as the law of value expresses the so cially necessary labor "to produce a new exemplar of the same commodity under generally given production-conditions," (Marx 1980b, 111; Marx and Engels 1975ff., Vol. 29, 273), the law of the production price is also a law of reproduction.

The same holds true for the average profit, the industrial profit, and the enterprise profit. Because they represent only quantitative modifications of surplus-value and because the latter enables the renewal of the production process in a capitalist form, therefore the profit as well "is the precondition for the fact that the new creation of means of production takes place in the *form of capitalist accumulation*! Thus, a relation dominating reproduction . . ." (Marx 1992, 899; Marx 1967, Vol. 3, 882).

The conclusion to be drawn, therefore is that the *fourth* goal of scientific explanation by gradual concretization, proceeding from a law of the $L_{ii}^{(k)}$ type, consists in a formulation of scientific laws explaining cyclical processes inside a natural kind due to which the latter endures and reproduces itself.

What is the picture like in the case of scientific explanation which proceeds from a law of the type $L^{(k)}$? Since this law is a concentrated expression of that type of scientific cognition which proceeds from the appearances of the essence towards the very essence, the cited cognition is unable to express the knowledge about essence independently from its appearances. Since the essence is defined here through the form of appearance $E^{(k)}$, the only possible basis for the explanation of the forms of manifestation $E^{(k-1)}, \ldots, E^{(0)}$ is not the essence H but the form of appearance $E^{(k)}$. Regarding the structure of the consequent in a law of the type $L^{(k)}$, the explanatory procedure has the form

$$E^{(k-j)} = f_{k-j}(E^{(k)}, Cmod_k, \ldots, Cmod_{k-(j-1)}) \tag{5.13}$$

When one compares this schema with (5.11) and (5.12), it becomes readily apparent that the effects of the impacts of modification conditions are here not correlated with the expression for the essence H but with the expression for the form of appearance $E^{(k)}$. This is the *first* difference between the course of scientific explanation by gradual concretization which proceeds from a law of the type $L_{ii}^{(k)}$ and that of explanation by gradual concretization which starts from a law of the type $L^{(k)}$.

The *second* difference manifested in the course of scientific explanation which proceeds from a law of the type $L_{ii}^{(k)}$ as opposed to one which starts from a law of the type $L^{(k)}$ is brought about by the different comprehension of the modification conditions. In scientific knowledge which hinges on a law of the type $L_{ii}^{(k)}$, the origin of the modification conditions is known, i.e., their functional dependence on the essence-producing substance is acknowledged. On the contrary, in a scientific law of the type $L^{(k)}$, their origin is unknown because the very essence-producing substance is unknown. Therefore, scientific explanation proceeding from a law of the type $L^{(k)}$ cannot arrive at any of the four aims reconstructed above. It cannot arrive at the first of the quoted aims since it contains no knowledge of the quality of the essence; neither the essential conditions nor the essence-producing substance are known. That is why it is impossible to explain the mechanism of the transfer of local fluctuations of the subkind's quantitative determinations into the entire natural kind if one begins with a scientific law of the type $L^{(k)}$.

The absence of these data, together with the failure to take into account that the modification conditions have been generated by the subkind's substance or generated by the interaction of the subkinds in one natural kind, prevents the above-mentioned explanation from achieving the second end, i.e., to explain the ties uniting subkinds into a functioning whole. Scientific laws of the type $L^{(k-1)}, \ldots, L^{(k-j)}$ $(k \geq j \geq 1)$, obtained from a law of the type $L^{(k)}$ by gradual concretization, explain the innate structure of the natural kind, its differentiation, only as a mere sum of mutually independent subkinds, of operationally incoherent instances of one natural kind, not as an active, operative whole made up of interconnections between these subkinds.

Since laws of type $L^{(k)}$ do not involve knowledge of the essence-producing substance, the explanation by gradual concretization which starts from it cannot reach the third cited aim. Scientific laws arrived at in this way do not explain the sources of the change in the quantitative determination of, nor in the quantitative proportions between, subkinds in a natural kind.

Since laws of type $L^{(k)}$ do not grasp the very essence, they cannot grasp the essence-producing process as a reproduction process. Therefore, the gradual concretization proceeding from them cannot reach the fourth reconstructed aim mentioned in the list above. Laws arrived at in such a way do not explain the internal structure of a natural kind as a structure reproducing itself.

Having reconstructed the difference between the explanation by gradual concretization proceeding from a law of the type $L^{(k)}$, which makes its appearance in classical mechanics, and the explanation by gradual concretization proceeding from a law of the type $L_{ii}^{(k)}$ appearing in the manuscript of the third volume of *Capital*, I arrive at the conclusion that Marx is the Galileo of the social sciences neither in terms of the type of scientific law employed, nor in terms of explanatory methods used in the manuscript of the third volume of *Capital*. What remains to be determined is whether Marx, as Nowak claims, was the first to introduce idealizations into the study of political economy. To answer this question one has to research carefully and in detail the history of pre-Marxian political economy.

5.4 THE DIALECTIC AS A LOGICAL METHOD OF SCIENTIFIC EXPLANATION AND AS AN HISTORICAL METHOD OF SCIENTIFIC EXPLANATION

Now it is possible to sum up the development of scientific knowledge through its various stages. It starts from 1 to l phenomena as appearances, and moves on towards the formulation of an empirical type of a scientific law. The latter already provides the basis for the simplest type of explanation reconstructed in the D-N model. Scientific laws of this type as well as explanations by deductive subsumption do not represent the *ultima ratio* of all science. Scientific knowledge transforms them into scientific laws which already exemplify the pure idealizational type of a scientific law. Explanation by means of gradual concretization rests on this foundation. Again, this type of scientific law as well as this method of scientific explanation do not represent the *ultima ratio*. The next step aims at the formulation of a law, exemplifying an inherent type of a scientific law, and expressing the basic and most simple relation – the "cell" – in the object under scrutiny. Scientific knowledge proceeds from it by explanation to the inherent type of a scientific law grasping the essence of the object under scrutiny. Then, scientific knowledge moves in two directions. Firstly, on the basis of the inherent type of a scientific law, it is necessary then to formulate laws corresponding to the inherent organic, the developmental, and the historical type of a scientific law. Secondly, on the basis of the inherent type of a scientific law, it is necessary to reconstruct how the essence (and its substance) becomes the causal ground for the manifestation of essence. This reconstruction materializes in the formulation of laws corresponding by their structure to the scientific inherent idealizational law of the kth degree of idealization. The latter then becomes the explanatory basis for obtaining inherent idealizational laws of the kth, $k - 1, \ldots, 0$ degree of idealization. The final step is the derivation of 1 to l phenomena, but now already as manifestations, by the introduction of the individual determinations for the variables appearing in scientific laws of the type $L_{ii}^{(0)}$. Marx made the following claim about the circular movement of cognition from the phenomena as appearances to the phenomena as manifestations with respect to the relation of the definition of value to other economic categories (1970, 315; Marx–Engels 1975ff., Vol. 40, 300–301):

> All objections to this definition of value . . . are based on the confusion whereby the more concrete economic definitions from which value has been abstracted – and which may therefore also be seen, on the other hand, as a further development of the same – are upheld as against value in this its abstract, undeveloped form.

This movement can therefore be labeled the movement of cognition from the sense concrete *via* the scientific abstract to the thought concrete. The 1 to l phenomena as appearances have the status of *senso-concrete phenomena*, P_{sc}^{1-l}, while the 1 to l phenomena as manifestations have the status of *thought-concrete phenomena*, P_{tc}^{1-l}.

At least two cases of such a circular movement in Marx's economical works are worth mentioning. The first pertains to the concept of commodity, which was chosen by Marx as a starting point for his economical considerations. He claims at the beginning of these considerations that "[t]he wealth of those societies in which the capitalist mode of production prevails, presents itself as an 'immense accumulation of commodities"' (1990, 29; 1967, Vol. 1, 35; Marx and Engels 1975ff., Vol. 35, 45). He claims also that he can investigate that accumulation of commodities by choosing an individual commodity as an "average sample of its class," (1990, 33; 1967, Vol. 1, 39; 1975ff., Vol. 35, 49) and as an "*universal elementary form of wealth*" (1988b, 28; 1975ff., Vol. 34, 359). The concept of commodity is for Marx a certain kind of an "observation-point" (Backhaus and Reichelt 1995, 108) by means of which the "observer" (a political economist) can start his proceeding in mind. But Marx starts here – at the very beginning of the first volume of *Capital* – from a point whose justification has as yet to be given exactly by this proceeding. He acknowledges this fact by explicitly speaking about the "circular course taken by our presentation" (1988b, 24; Marx and Engels 1975ff., Vol. 34, 355). He starts from the fact of the existence of an immense accumulation of commodities, and then, *via* the concept of surplus-value, explains in "Sechstes Kapitel. Resultate des unmittelbaren Produktionsprozesses" (1988b, 24–130; 1975ff., Vol. 34, 355–466) this accumulation. He claims as follows (1988b, 33; 1975ff., Vol. 34, 363)

> [t]he *commodity* as it comes out of the capitalist production differs in its determinations from the commodity from which one proceeds as an element, presupposition of the capitalist production. We proceeded from the individual (einzelne) commodity as an independent article in which a certain amount (Quantum) of labor-time has been objectified . . .
>
> The commodity appears now as twice determined:
>
> 1) What is objectified in it . . . is a certain amount of socially necessary labor. But while in the case of the individual commodity it was entirely undetermined . . . from whom this objectified labor originated, etc., the *commodity* as the as the product of capital contains partly paid, partly unpaid labor . . .
>
> 2) The individual commodity appears not only materially as part of the total product of the capital, as an aliquot part of the lot produced by it. We have not at all in front of us the individual commodity, the individual product. Not singular commodities appear as the result of the process, but a *mass of commodities*, in which the value of the advanced capital + surplus-labor . . . has reproduced itself.

This means that by explaining the production of this whole mass of commodities, Marx derives exactly that point from which he initially departed in the investigation of use-value and exchange-value.

The second case of a circular movement pertains to Marx's derivation/explanation, in the manuscript of the third volume of *Capital*, of the initial supposition from its first volume, that a certain chosen commodity can be viewed as an "average sample" (1990, 33; 1967, Vol. 1, 39; Marx and Engels 1975ff., Vol. 35, 49) of the whole class of commodities (i.e., produced, from the quantitative point of view) by a certain amount of socially necessary labor. Neither in the first nor in (the manuscripts of) the second volume of *Capital* does Marx explain how that average comes into being. In the third "subchapter" ("Equalization of the General Rate of Profit Through Competition") of the second chapter of the manuscript of third volume he states that (1992, 256; 1967, Vol. 3, 181)

> [l]ooking closer, we see that the conditions that obtain for the *value of the individual commodity* reproduce themselves here as conditions determining the *value of the total amount of a certain kind of commodity*; capitalist production is, from the very outset, mass production . . .

and that (1992, 257; 1967, Vol. 3, 182)

> [w]hat we said of the individual commodity now applies word for word to the mass of commodities of a certain branch which is on the market. The requirement that the individual value of a commodity should correspond to its social value is now realized, or further determined, in the sense that the total mass contains the socially necessary labor necessary for its production, and that value of this mass = its *market value*.

Only then does he explain that average under the supposition that the branch of production under consideration turns out just one type (class) of commodities. He presupposes also that this branch has three groups of production processes: normal; better than normal; and worse than normal; the last two representing – with respect to the first – extreme conditions of production. He then proves that the value-determining socially necessary labor-time required for the production of one commodity from this branch is determined by the average of the real labor-times in the different production processes of this branch (for a mathematical treatment see Quaas 1988). While this pertains only to one branch of production and the market-value of the commodities in it, the same can be stated, according to Marx, about the whole economy and the production prices of commodities produced in it.

The movement of thought from the scientific abstract to the thought concrete represents a succession of different types of scientific explanations which can be symbolized as follows ("$=_{tr}=|$" denotes transformation, "$-_{gc}\dashv$" denotes gradual concretization, "$\supset$" denotes deduction, "$\Rightarrow$" stands for the grasping of the genesis of the manifestation of the immanent measure; it will be analyzed in subchapter 6.4).

$$L_i^A \,\&\, Cim_{1-m}^B =_{tr} = |L_i^{AB} \Rightarrow L_{ii}^{AB(k)} -_{gc}\dashv L_{ii}^{AB(k-1)} \ldots -_{gc}\dashv L_{ii}^{AB(k-j)} \,\&\, Ind_{1-l} \supset P_{tc}^{AB(k-j)} \tag{5.14}$$

The thought move from a law of the type L_i^A towards a law of the type L_i^{AB} is an explanatory move from the most elementary form of natural kind under investigation – the "cell" – towards the essence, where the latter is a structure which has historically (genetically) become, i.e., a genetic structure formed by a unity of two inherent laws. Such a genetic structure is exemplified by the law of surplus-value in the first volume of *Capital*. Surplus-value is here understood as an essentially new determination of value – value created by unpaid labor – i.e., surplus-labor. The method of scientific explanation by means of which one obtains the law of essence of a genetic structure is called the structuro-genetic method of scientific explanation, and it is exemplified by Marx's move from value in Part I to surplus-value in Part III of the first volume of *Capital*. The explanatory move from L_i^A up to the phenomena as manifestations represents dialectics as a logical method of scientific explanation. Its aim is to reconstruct the internal structure of one natural kind. Dialectics as a logical method of scientific explanation is exemplified by Marx's move from the law of value and surplus-value in the first volume of *Capital* to the categories like production-price, market-value, market-price, etc. in the manuscript of its third volume.

Let me now analyze the movement of cognition from the scientific developmental law towards a scientific historical law. This movement is based on the idea of a thought-prolongation of the non-simple reproduction process theoretically grasped in a scientific law of the type L_d. This idea was applied by Marx in his reconstruction of the historical genesis of surplus-value and capital (1990, 619; 1967, Vol. 1, 713; Marx and Engels 1975ff., Vol. 35, 704) as follows:

> We have seen how money is changed into capital; how through capital surplus-value is made, and from surplus-value more capital. But the accumulation of capital presupposes surplus-value; surplus-value presupposes capitalist production; capitalist production presupposes the pre-existence of considerable masses of capital and labor power . . . The whole movement, therefore, seems to turn in a vicious circle, out of which we can only escape by supposing a primitive accumulation . . . preceding capitalist production; an accumulation not the result of the capitalist mode of production, but its starting point.

By such a thought-prolongation one arrives at "the points where a historical analysis must be introduced" (Marx 1981a, 369; Marx and Engels 1975ff., Vol. 28, 388). On the one side one finds points marking the historical becoming of the essential conditions and therefore of the very essence, i.e., when the scientific law of the type L_h points at "the past lying behind this system," (Marx 1981a, 369; Marx and Engels 1975ff., Vol. 28, 389) and on the other side one finds points marking the historical decline of the essential conditions and of the very essence. Here the scientific law of the type L_h already points to "foreshadowing of the future, of the becoming motion" (Marx 1981a, 369; Marx and Engels 1975ff., Vol. 28, 389). This first aspect of the thought-prolongation means that the type L_h of a scientific law, formulated for a certain natural kind, becomes a key for the theoretical comprehension of the essence of the natural kind from which that kind really historically developed. This fact is expressed in the notorious phrase of Marx "the anatomy of the man is a key to the anatomy of the ape" (1976, 40; Marx and Engels 1975ff., Vol. 28, 42). According to him when one theoretically comprehends the relations of the capitalist mode of production, then this comprehension also provides the key for understanding "the structure (*Gliederung*) and relations of production of all those declined societal forms, with ruins and elements of which they have built themselves up" (Marx 1976, 40; Marx and Engels 1975ff., Vol. 28, 42). If, for example, we replace in the scientific law of surplus-value the essential condition Cim_1x by a condition stating that x is in the position of a subject (with respect to y as a lord), not a free owner of his labor power; the essential condition Cim_3x,y by the condition that y supplies the means of production directly to x, and therefore can directly usurp alien work; the essential condition Cim_4y,x by the condition that y views as the ultimate goal of his relation to x the enjoyment or creation of use-value, then we reduce the scientific law of surplus-value to a scientific law expressing the appropriation of the surplus-product in the pre-capitalist societies.

So, generally, the attempt here is to grasp the structure of certain natural kinds and grasp their historical becoming through knowledge about a natural kind which emerged from them. But this means that real history is theoretically grasped here only insofar as it has led to this natural kind. From the determinations of the historically previous natural kinds, one reconstructs here only what was absorbed and preserved inside the later natural kind, which is a point of departure of scientific knowledge. The structuro-genetic method of explanation for the natural kind's essence can provide the key for successive theoretical reconstructions of the essences of historically previous natural kinds, but it cannot replace their anatomies. Marx's claim that the anatomy of man is a key to the anatomy of the ape needs one amendment, namely, "but it cannot replace this anatomy." This fact is readily seen when one takes this method as a key for the reconstruction of the real transformation process between two natural kinds and for the comprehension of the structure of the historically earlier natural kind. One might be tempted to draw a conclusion which can be symbolically represented as follows:

$$L_{i*}^{A} \,\&\, Cim_{1-r}^{B} =_{tr}= |L_i^{AB}$$
$$L_{i}^{A} \,\&\, Cim_{1-r}^{B} =_{htr}= |L_i^{AB}$$

The movement from a scientific law of the "cell," L_{i*}^{A}, towards the essence,[4] L_i^{AB}, inside one natural kind by means of the essential conditions Cim_{1-r}^{B} could seemingly found the following claim. The essence, L_i^{AB}, of a natural kind originated in the historical transformation process, $=_{htr}=|$, where the historically created conditions transformed the law L_i^{A} representing the essence of the historically prior kind. But such a move is based on a false move in thinking, namely it is false to measure the order in the genetic structure of historically previous natural kinds by means of the order in the genetic structure of the present natural kind with the law of essence L_i^{AB}. As a result of such a move the order in the genetic structures of historically previous natural kinds would be understood as in principle identical with the order in the genetic structure of the natural kind with the law of essence L_i^{AB}. The structuro-genetic method contains the historical (genetic) moment as an indispensable, yet subordinated, moment. The former's ultimate aim is the theoretical reconstruction of the structure, functioning, and development of the essence of one and only one formed natural kind functioning, moving, and developing already on its own ground. It cannot, therefore, grasp either the order in the genetic structure of historically previous natural kinds, nor can it grasp the real historical process which is a sequence of natural kinds with different essences developing one from another. It is, therefore, necessary to reflect on *dialectics as a historical method of scientific explanation.* Its starting point is the movement from the knowledge expressed in L_{i*}^{A}, Cim_{1-r}^{B}, and L_i^{AB} towards ${}_hCim_{1-r}^{B}$, as well as towards the structure containing the inherent law L_i^{A}, which can be termed "historical retrospective." The knowledge of the essential conditions of type Cim_{1-r}^{B} as well as of the law of the "cell" serves as a key for sorting out the relevant historical facts. So, one can pose the following questions: "In what types of historical processes did these essential conditions really originate?" and "What was the genetic structure of the historically previous natural kind?" What should follow then is a full reconstruction of the transformation process of the previous natural kind as well as a complete reconstruction of the order in its genetic structure. In symbols:

$$L_{i*}^{A} \,\&\, Cim_{1-r}^{B} =_{tr}= |L_i^{AB} \tag{5.15}$$
$$\downarrow \qquad \downarrow$$
$$L_i^{AD} \,\&\, {}_hCim^{D} \Rightarrow 0 \,\&\, {}_hCim_{1-r}^{B} =_{htr}= |L_i^{AB} \tag{5.15'}$$

In (5.15′) real history is theoretically grasped as a process of the historical decline of the essence L_i^{AD} and its historical transformation into essence L_i^{AB} of a new natural kind through the historical decline of the essential conditions Cim^D and the historical becoming of the essential conditions Cim^B. Formula (5.15') symbolizes the historico-genetic method of scientific explanation. It presupposes, as a subordinated moment, the structuro-genetic method of scientific explanation, symbolized by (5.15), as well as the thought-move labeled "the historical retrospective," symbolized by the vertical arrows. In subchapter 11.3 I will show that dialectics as logical method of scientific explanation represents operations with the *ideal average type of a thought-object* (Marx 1992, 853; Marx 1967, Vol. 3, 831), while the dialectics as historical method of explanation represents operations with the *original type of a thought-object* (Marx 1980b, 113; Marx and Engels 1975ff., Vol. 29, 275).

5.5 THE INHERENT TYPES OF SCIENTIFIC LAW AND THE POZNAŃ SCHOOL

In my view, a scientific law of the type L_i and a scientific law of the type $L_{ii}^{(k)}$ are more developed than a scientific law of the type $L^{(k)}$. In the former, contrary to the latter, the essence is grasped independently from appearances and manifestations. On this basis it is then possible to theoretically reconstruct the becoming of the manifestation $E^{(k)}$, and also therefore of the manifestations $E^{(k-1)}$, . . ., $E^{(0)}$. In a scientific law of the type $L_{ii}^{(k)}$ the essence-cause is the reason and the effect is the reasoned; the contradiction between the cognition of the cause-effect relation and the real genesis of the cause-effect relationship, typical for laws of the type $L^{(k)}$, is already overcome here. On the basis of the concept of inherent law and inherent idealizational law, it is possible to reconstruct methods of scientific explanation which are more advanced with respect to the method of scientific explanation proceeding from laws of the type $L^{(k)}$.

If one compares Nowak's reconstruction of the scientific idealizational law as expressed in (3.2) with our reconstruction expressed in (5.3), (5.4) and (5.4'), it is readily seen that his approach represents an erroneous "synthesis" of the antecedent and of the symbol for the sentential connective from $L^{(k)}$ with the consequent of $L_{ii}^{(k)}$. How could such a "synthesis" have arisen at all? If we take a closer look at Nowak's reasoning concerning the scientific law of value and surplus-value, it is evident that he has mounted this reconstruction on the basis of the knowledge of the third volume of *Capital*, specifically on the basis of the knowledge of forms of manifestation derived by Marx (market price, production price, interest, profit, rent). Having become aware of the significance of the modification conditions for scientific laws, he introduced idealizations and so retroactively reduced these forms of manifestations to their foundation – to value and surplus-value (i.e., to the starting points from which Marx derived them). Nowak unwittingly uses a law of the type $L_{ii}^{(k)}$ as his starting point, and it contains the equation with the structure $E^{(k)} = g_k(S)$. On the basis of these reflections, Nowak became aware of the consequent with the structure $E^{(k)} = f_k(H)$. Simultaneously, Nowak became aware of the significance of idealizations for physics. He reconstructed the antecedent of the scientific laws of physics which, indeed, involve both the name of a natural kind and idealizational conditions. This of necessity led him to the formulation of the antecedent in the form $N \,\&\, Cmod_{1-k} = d_{1-k}$. His understanding of the structure of the idealizational law expressed in (3.2) grew out of a "synthesis" of the antecedent thus reconstructed and the consequent. My analysis, I hope, has shown that the antecedent and the consequent do not match each other. The antecedent, which is correct with respect to classical mechanics, requires a different consequent. The consequent, which is correct with respect to the manuscript of the third volume of *Capital*, requires another antecedent.

As a result of his failure to distinguish between the two types of idealizational law, $L^{(k)}$ and $L_{ii}^{(k)}$, Nowak could not distinguish between the two methods of scientific explanation by gradual concretization.

Despite all these drawbacks, the philosophy of science created by the Poznań School, and especially by Leszek Nowak, represents a tremendous achievement. Firstly, in contrast with the theories proposed by K. R. Popper and C. G. Hempel, the theories put forward by the Poznań School arrived at the concept of the modification conditions. Secondly, the work of these scholars introduced the model of scientific explanation by gradual concretization, thus pointing the way to a solution of Hempel's Footnote 33 problem. Thirdly, this work paved the way (even if unwittingly) for the reconstruction of the missing link between the empirical type of scientific law and the inherent type of scientific law.

CHAPTER 6

THE THREE FACES OF MEASURE

In subchapter 3.5, while distinguishing the pure idealizational type of a scientific law (exemplified by Newton's second law of dynamics) from the type of scientific law exemplified by the law of value and the law of surplus-value, I used the category of the formal ground as well as the category of the real ground. In subchapter 5.4, when analyzing the development of knowledge from the inherent type of scientific law to the inherent idealizational type, I characterized this development as a move from the category of immanent measure to that of the manifestation of the immanent measure. The task of this chapter is to show the interdependence of these categories and to relate them with the category of apparent (external) measure.

6.1 ELLIS ON FORCE: THE FORMAL AND THE REAL GROUND

The category of the formal ground was explicated in subchapter 3.5 by means of the concept of force as given in classical mechanics. I start here from this concept.

An excellent treatment of this concept by means of the category of measure was given by B. Ellis in (1965). The principal points of Ellis's paper are as follows:

1) The existence of forces is both entailed by and entails the existence of effects they are supposed to produce (1965, 31).
2) If the second dynamic law, $F = ma$,[1] is true by definition, then the concept of force lacks any explanatory power because one cannot explain the existence of an effect by postulating the existence of something whose sole *raison d'être* is that it produces the given effect (1965, 54).
3) The product of the mass and acceleration of an object defines a scale for the measurement of the resultant force acting on that object (1965, 54).
4) Why should this scale and not another be taken as a scale of force? What is the relation between scale and force; is it analytic or synthetic (i.e., empirically discovered) (1965, 54)?
5) Having in mind point 2 above (i.e., that the rate of change of momentum should not be identified with force) one faces yet another question. How can the measure of the resultant force acting on a given system be identified with the measure of the rate of change of the momentum?
6) The two questions posed in points 4 and 5 above are related to issues of scales and quantities:
   i) the existence of a quantity depends upon the existence of an objective linear order. For if the objects (systems, events, states, etc.) possess a given quantity q, it is always possible to arrange those objects in order of q by some objective ordering procedure. An objective ordering procedure is any which, if perfectly executed, would always lead to the same ordering among the same particulars under the same conditions (1965, 53–54);

ii) if two or more logically independent and objective ordering procedures would always in fact generate the same order among the same particulars under the same conditions, then we should say that they are procedures for ordering those particulars with respect to quantity q (1965, 55);
iii) there are many, logically independent procedures for ordering things in respect of the same quantity;
iv) it is the order, not the ordering relationship, which provides science with the criteria for the identity of quantities. This order may be identified by any number of logically independent ordering procedures.

7) But from 6i–iv it does *not* follow that any particular ordering procedure is essential to the concept of any quantity. So, as we will see below, Newton in the *Principia* distinguishes between the accelerative and the motive force. Similarly Ch. Huyghens (see on this (Jammer 1957, 109–112)) used the distance covered by a body as a measure to determine the force acting on it in a short interval of time. So, mechanics has at least three procedures to determine the quantity called "force": by acceleration, by change of motion, and by the covered distance. Force is, therefore, a cluster concept. There are various procedures which can be used to determine the quantity labeled "force"; none of them has the status of a necessary one.
8) The claim, according to Ellis, that there is one universally applicable, necessary procedure to determine the order of a quantity q constitutes an acceptance of the Lockean doctrine of real essences. "But," according to Ellis, "most philosophers nowadays would reject this doctrine for the very good reason that they have now become clear as to the 'cluster' nature of many of our concepts" (1965, 59).

Ellis's approach, in my view, is the expression of the real situation one encounters in classical mechanics when studying its concept of force. The aim in classical mechanics, as I have shown in Chapter 3, is to proceed from phenomena as appearances to their ground (forces), and then from the latter to derive or explain the phenomena as manifestations. Classical mechanics, however, suffers from the problem that it does not state what force is in itself. This epistemic fact, the fact that cognition in the special sciences has to pass through a level where the ground is initially defined only by means of the appearances, was not realized by Ellis. This becomes apparent when he speaks about force, for example, and uses the term "quantity," thereby covering two connected, but nevertheless different, aspects of force.

But force should be understood as a unity of quantity and quality. If I say that a body with a mass of five kilograms, *5 kg*, was accelerated to four meters per second per second, $4\ ms^{-2}$, by a force of twenty Newtons, *20 N*, then the latter unifies in itself a quantitative aspect, symbolized by the numeral "20" and a qualitative aspect, symbolized by the letter "N". Classical mechanics has no problem with determining the quantity of force; here it is twenty of something. But it faces a serious problem when it comes to defining this something with respect to quality (i.e., what is force as a quality?). So a more precise term, which I will use instead of Ellis's "quantity," is "magnitude." The quantitative aspect of a magnitude q is usually expressed as $\{q\}$, while its qualitative aspect is expressed as *[q]*. The magnitude *q* as a unity of both these aspects is usually symbolized as $q = \{q\} \cdot [q]$, where "·" stands for multiplication.

The category of the formal ground and the category of the real ground were introduced by Hegel in the *Science of Logic*. It is noteworthy that Hegel, by clearly distinguishing between them, explicitly and consciously addressed problems which were, in an unacknowledged way, expressed 150 years later by Ellis. Hegel's views can be summed up as follows:

1) at the level of cognition which is based on the knowledge of the formal ground, the same content is given for the ground which was previously given for the appearances; (1923, T. 2, 79; 1969, 459)[2]

> For example, the ground of the movement of the planets round the sun is said to be the *attractive* force of the earth and the sun on one another. As regards content, this expresses nothing other than what is contained in the phenomenon, namely the relation of these bodies to one another . . . If one asks what kind of a force the attractive force is, the answer is that it is the force that makes the earth move round the sun; that is, has precisely the same content as the phenomenon of which it supposed to be the ground; the relation of the earth and sun in respect of motion is the identical basis (*Grundlage*) of the ground and the grounded.

What Hegel has in mind can be expressed by means of the pair of ontological categories "ground" and "grounded by," as introduced already in subchapter 3.5. To escape the problem (in the words of Ellis) that force in mechanics is both entailed by and entails the effects it is supposed to produce, then physics has to provide the knowledge of two different bases (Hegel's *Grundlage* in the above given quotation): one for the very force, and another one for its effects.

2) According to Hegel, the effects are, on the one hand, the reasons for postulating the existence of a force but, on the other hand, they are grounded by force (1923, T. 2, 79; 1969, 458).
3) In mechanics one has to find in the phenomena the reasons for the ground to be derived, "instead of being in and for itself and self-subsistent," and the ground is here "organized (*eingerichtet*) according to the phenomenon and its determinations are based on it" (1923, T. 2, 80; 1969, 459).
4) Further, in mechanics the ground is, (1923, T. 2, 80; 1969, 459)

> on the one hand, ground as the reflection-into-self of the determination of the content of the phenomenon which it grounds, on the other hand, it is the posited. It is that from which the phenomenon is to be understood; but *conversely*, it is the ground that is inferred from the phenomenon and the former is understood from the latter . . .

From all these ideas, Hegel, having in mind that the mechanics of his time did not know the internal essence of forces (1923, T. 2, 82; 1969, 461), draws two conclusions. Firstly, he claims that in mechanics "cognition has not got from the spot (*vom Flecke gekommenn*)" (1923, T. 2, 80, 1969, 459). But this holds only partially. By finding out that the acceleration $4\ ms^{-2}$ of a body with a mass of *5 kg* was caused by a force of *20 N*, one discovers a certain quantity. Cognition, which started with the knowledge of the quantity 4 and 5, ends up with the knowledge of the quantity 20. Of course, from the point of view of quality, cognition has not gone a single step further;

no growth of knowledge took place. The quality, expressed by the symbol "N," is "defined" by the qualities of the phenomena from which cognition started. It started from the knowledge of the quality of distance, *[l]*, time, *[t]*, and mass, *[m]*, and ended up with the knowledge of the quality of force, so that $[f]:=[m]\cdot[l]\cdot[t]^{-2}$ holds, or, by means of the system SI, $[N]:=[kg]\cdot[met]\cdot[sec]^{-2}$.[3] Nothing new was discovered; thus, the quality of the ground is given only as a "multiplication" of the qualities of its effects.

Secondly, Hegel recommends that mechanics should either find something else which constitutes force or stop at simple facts (1923, T. 2, 82; 1969, 461). But in my view, there is no something else. Physics has to pass through this level of cognition (i.e., the level of knowledge of the appearances) but it should not stop there.

6.2 FORCE IN CLASSICAL MECHANICS

It is crucial to analyze how classical mechanics confronts those questions of force which stem from the fact that force is understood only as a formal ground. As an empirical basis of my analysis I take Newton's *Principia*. I focus here on Newton's ambitions and will compare the targets he explicitly set with the goals he actually attained.

The goal pursued by the *Principia* can, in my view, be understood as an attempt to unify the knowledge of the various physical phenomena according to a single basis, and then explain them on this basis (i.e., to transform them into manifestations of this ground). In fact, Newton, states this aim in the introduction to the first edition of the *Principia* (1946, XVII–XVIII; 1687, A2):

> [T]he whole burden of philosophy seems to consist in this – from the phenomena of motion to investigate the forces of nature, and then from these forces to demonstrate the other phenomena; and to this end the general propositions in the first and second Book are directed. In the third Book I give an example of this in the explication of the System of the World; for by the propositions mathematically demonstrated in the former Books, in the third I derive from the celestial phenomena the forces of gravity with which bodies tend to the sun and the several planets. Then from these forces, by other propositions which are also mathematical, I deduce the motions of the planets, the comets, the moon, and the sea. I wish we could derive the rest of the phenomena of Nature by the same kind of reasoning from mechanical principles, for I am induced by many reasons to suspect that they may all depend upon certain forces . . .

Let me, therefore, analyze Newton's approach to the concept of force. Force is for him the ground upon which physics should base its attempts to unify and explain the various phenomena of nature. The concept of force (*vis*) appears for the first time in the *Principia* in Definition III, where he views inertia as a force inherent to bodies (*vis insita*). Definition IV introduces the notion of impressed force; according to him "of this sort is gravity" (1946, 2; 1687, B2). Of crucial importance for this chapter are Definitions VI, VII and VIII (1946, 4; 1687, B2):

> [i] The absolute quantity of a centripetal force is the measure of the same, proportional to the efficacy of the cause that propagates it from the center, through the spaces round about.

> [ii] The accelerative quantity of a centripetal force is the measure of the same, proportional to the velocity which it generates in a given time.

[iii] The motive quantity of a centripetal force is the measure of the same, proportional to the motion which it generates in a given time.

Then, for the sake of brevity, he labels these magnitudes as the *absolute*, *accelerative*, and *motive* forces.

Let me analyze the last two definitions. What unifies them is the fact that in both the measure of the force is determined by the measure of the effects of this force, those effects being change of velocity in a given time and change of motion in a given time. In other words, in these definitions the quantitative and qualitative determinations of the effects – their measure – are used in cognition to determine the quantity and quality of their ground-cause. These measures are, with respect to force, only its apparent (external) measures. Now what about the internal, inherent (immanent), absolute measure of the very force as it exists in itself, independently of phenomena like change of velocity or change of motion? In Definition VI Newton explicitly mentions the absolute force and in the commentary to definition VIII he characterizes it as a "cause . . . that does not yet appear" (1946, 5; 1687, [4]). But here he does not say what its internal measure is. In Definition VII and VIII something different from force – namely something external to it (change of velocity or change of motion in a given time) – is related to force as its measure. But neither in Definition VI nor elsewhere in the *Principia* does one find any reference to something different from force – something that would be internal with respect to force, i.e., related to it as its measure. This fact is not surprising at all, for Newton, in the very introduction to the first edition of the *Principia,* already states, with respect to the movement of planets, comets, and the moon, that[4] (1946, XVIII; 1687, A2):

they may all depend upon certain forces . . . These forces being unknown, philosophers have hitherto attempted the search of Nature in vain; but I hope the principles here laid down will afford some light either to this or some truer method of philosophy.

So, the first moral of the "epistemic story" exemplified by the *Principia* is as follows. If scientific cognition, in its attempt to move from phenomena-effects as appearances of a ground-cause to the phenomena-effects as the latter's manifestations by means of the category of ground-cause, temporarily fails to grasp what this very ground is (i.e., its internal measure), then the knowledge of this ground is characterized not only by the category of the formal ground, as claimed in Chapter 3, but also by the category of apparent (external) measure.

The reconstruction of the type of scientific knowledge which is based on the category of the formal ground and apparent measure enables me to find a solution to Ellis's problem given under number two in subchapter 6.1. In order to explain certain effects within the confines of such a type of knowledge one is compelled to postulate the existence of something whose sole *raison d'être* is that it produces these effects. At this level, however, there is as yet no knowledge that can identify the ground's *raison d'être* as something which is independent from its effects. But, contrary to Ellis's claim, the knowledge of the ground as a formal ground does not lack any explanatory power. It yields the knowledge of the quantitative determinations of the ground. This knowledge, in turn, can serve as the basis for the derivation of the quantitative determinations of other effects of this ground. As shown above in subchapter 3.3, we can determine the quantity (size) of the period of a simple pendulum or the path covered by a free-falling mass-point, both under the impact of the force F, where for the latter holds $F = mg$. Similarly, by deriving the law of gravitation from the knowledge of the path covered by a planet moving around its sun (Kepler's laws) and from the law $F = ma$, we can determine the quantity (size) of the force of gravity. This, in turn, enables us to determine the quantity of the effects of the gravity of this planet on a body, say, the quantity of circular velocity of this body around this planet in a certain distance from it. Knowing this quantity we can, then, determine the three cosmic velocities of this body: cosmic velocity one this body has to reach in order to move at the same circular speed as the circular velocity of the planet from which it was launched; cosmic velocity two it has to reach in order to escape the gravitational field of this planet; cosmic velocity three it has to reach in order to escape the gravitational field of that planet's sun.

The type of scientific knowledge characterized by the categories of the formal ground and of the apparent (external) measure has very specific and peculiar characteristics, which can be explained by turning again to the *Principia*. On the one hand, Newton distinguishes the absolute force as cause from its effects (change of velocity, change of motion). With respect to the case of gravity as a centripetal force, he claims in proposition seven of the third book that "there is a power of gravity pertaining to all bodies, proportional to the several quantities of matter which they contain" (1946, 414; 1687, Bb2). This means, if taken together with Definition VI cited above, that the absolute quantity of the gravitational force depends on the mass of the central body. But, on the other hand, Newton in the commentary on the third rule of reasoning (Book III) in the second edition of the *Principia* makes a different claim[5] (1946, 400):

Not that I affirm gravity to be essential to bodies: by their *vis insita* I mean nothing but their inertia. This is immutable. Their gravity is diminished as they recede from the earth.

These views, if taken together, reveal that Newton did not always distinguish properly between gravitation as the ground-cause-essence and one of its effects: mutual attraction.

This muddling of the distinction between the ground and its effects which becomes evident in scientific cognition of the type that knows the ground of the objects it scrutinizes only as a formal ground, was described by Hegel as follows (1923, T. 2., 81; 1969, 461):

In such an exposition, therefore, one does not know where the grounding is the derived, nor (*worin das eigentlich Begründende als Abgeleitetes vorkommt, nicht, weder*) does one know how to take either ground or phenomenon.

The second moral which can be extrapolated here is that the categories of the formal ground and of the external measure, even if sufficient for the scientific cognition proceeding from the phenomena as appearances to the ground grasped as a formal ground, are not sufficient for grasping the latter as a real ground, a movement that would enable one to discover the immanent measure of the cause of those effects. One deals here with two stages of cognition, and these differ from each other by their categorial "outfit," by their cognitive depth. The knowledge of the phenomena suffices for grasping the ground as formal. But in order to grasp it as a real ground, i.e., as it is in itself, independent of appearances (and prior to the derivation of its manifestations), one has to abstract from all phenomena and has to move in thought in the "pure" essence. As shown already in chapter five, the categories of thinking which move in the "pure" essence of an object of a certain natural kind are those of essential conditions, reproduction of essential conditions, essence-producing substance, and, as shown in this subchapter, that of inherent (immanent) measure. What, in general, are categories of thinking? This will be explained in chapter ten.

It seems to me that Newton did realize that the framework in which he deduces forces from the phenomena of motion is not a framework in which one can grasp what those very forces are. This realization is expressed in two ways in the second edition (1713) of the *Principia*. Firstly, at the very beginning of his discussion Newton clearly differentiates between hypotheses which are of a transphenomenal character (*Rules of Reasoning in Philosophy*) and hypotheses derived from astronomical phenomena (*Phenomenon I* through *Phenomenon VI*) which are subject to observation and calculation. Secondly, he is well aware of the fact that his approach to force yields the understanding of its quantity, but leaves completely open the issue of its quality. This is expressed by Newton in the claim that the purpose of the investigation of force (1946, 550)

> is only to trace out the quantity and properties of this force from the phenomena . . . , and to apply what we discover in some simple cases as principles, by which, in a mathematical way, we may estimate the effects thereof in more involved cases; . . . We said *in a mathematical way*, to avoid all questions about the nature or quality of this force, which we would not be understood to determine by any hypothesis . . .

6.3 HEGEL ON MEASURE

The first philosopher in the European tradition who dealt at length and in detail with the category of measure was Hegel. In the first two chapters of the categorial cluster "Measure" from the book *Being* of the *Science of Logic*, he reconstructed categories of thinking which are embedded in this cluster. According to him, thinking starts by taking something. Say it is a piece of iron, and this has a determined quantum – a piece. One can state here that "all that exists has a measure" (1923, T. 1, 343; 1969, 333). Following a decision, a certain quantum is taken as a unit, let us say it is taken as a certain part of the earth's diameter. Using this quantum as a basis, one then establishes a standard (*Maßstab*). Up to this point, the knowledge of measure was only the knowledge of a *specific* quantum, but here it gains a rule (*Regel*). One then measures something by means of this standard and determines it as an amount (*Anzahl*) of this unit. If this specific quantum is related to another specific quantum, which is external to it, then one obtains the knowledge of the specifying measure (*das spezifizierende Maß*). Measure, according to Hegel, is here a relation of two sides: a relation of something external to something internal, say, the external temperature of the environment to the internal temperature of a body. The internal reacts to the quantitative changes of the external by means of specific internal changes; it transforms the external quantum into an internal one. These two sides are in a certain *ratio* (*Verhältnis*), so the measure is this ratio and its specificity is the exponent (*das Exponent*) of this ratio. This relation of the external and the internal, initially viewed as the exponent, where on both sides the same quality appears – one as a quantitative and the other as a qualifying (*qualifizierende*) – is in fact a ratio of two qualities. So, each side of the specifying measure is in itself a measure, and the whole is a measure given by the quantitative relation of these two qualities.

Here, according to Hegel, the notion of the *variable magnitude* (*veränderliche Grösse*) enters in. One side, that which changes externally, is a certain amount that increases or decreases and determines in a specific way the amount of the other side. The former is a unit (*Einheit*) with respect to the latter. Hegel makes reference here to two cases from mechanics: the determination of velocity by the ratio of the covered distance and the time elapsed, $v = s/t$, and the determination of the covered distance for the free fall, $s = at^2$. In the first case the unit is the divisor, while the amount is the dividend. In the second case, exemplifying the case of a specifying ratio, the amount is the power and the unit is the root. Hegel, by comparing these two ratios, assigns a higher cognitive value to the latter. In the former, either time or space can be chosen to represent either a unit or an amount. But in the latter case the time is the empirical quantum changing in an arithmetical progression, while the covered distance is a specified amount.[6]

The next level that thinking may reach is, for Hegel, the level of the category of being-for-self in measure. He takes the two sides of the specified measure. This measure provides cognition with the knowledge of a quantum which remains fixed with respect to all changes of that other side of the specifying measure which is the unit. Hegel takes as an example of this the ratio of the covered distance and elapsed time for the fall of bodies. He stresses that even if this quantum, acceleration, *a*, is derived by the specified measure, it is ascribed to something which is different from the ratio s/t^2; it is the force of gravity. Thus, because acceleration is derived from the specified measure, it is also a measure; a measure of the force of gravity. But as the knowledge of acceleration is given in the form of $a = s/t^2$, it is empirical knowledge (1969, 346). With respect to the very force of gravitation, it holds that the empirical quantum – acceleration – is, firstly, different from the very force of gravity. Secondly, as an empirical quantum it is, with respect to the force of gravity, its *empirical, external measure* (1923, T. 1, 356-357; 1969, 346):

> The said empirical quantum is attributed to the force of gravity, on the supposition that this force itself has no connection with the actual specification, with the ratio of powers characteristic of a determination of measure. The *immediate* moment, that in the motion of a falling body the amount of some fifteen spatial units . . . is traversed in a unit . . . of time, is an *immediate measure* . . . ; however, the concrete sciences have so far failed to throw any light on the basis of determination of such numbers.

What has to be stressed here is that Hegel strictly differentiates between the type of knowledge by means of which special science proceeds to the external measure and the type of knowledge that can discover what is really behind the external measure (i.e., a type of knowledge that can find whatever it is that this external measure points to). This is readily seen in the addendum to paragraph 270 of his *Encyclopaedia*. Here, in defense of Newton, he states (1842, 110–111; 1970a, 74–75; 1970b, Vol. 1, 272)

An injustice has been done to Newton insofar as gravity is also in mathematics understood in two different ways. In the first instance, it is simply the direction in which a stone falls, fifteen feet a second on the surface of the earth, and as such it is a purely empirical determination. Newton applied the law of falling bodies, which is ascribed primarily to gravity, to the revolution of the moon, for this also has the earth as its center. The magnitude of fifteen feet is thus also laid as a basis for the moon's revolution. Since the moon is at a distance of sixty earth-diameters from the earth, the moment of attraction in its movement is determined according to it. Afterwards it is found out that which produces the attraction of the earth on the moon . . . simultaneously determines the whole of revolution of the moon: it falls also . . . But this is initially only a singular case, the extension of the empirical fall on the earth to the moon . . . Therefore only the second meaning of gravity is universal gravitation, and Newton saw in gravity the law of all motion. He thus transferred gravity to the law of the celestial bodies, and named it the law of gravity.

Finally Hegel analyzes thinking which moves in the framework of the category of real measure. It starts from the ratio of independent measures (*das Verhältnis selbständiger Maße*). Measure becomes here a ratio of specified measures so that a whole, which is the ratio of them as they relate to one another, is initially just an external connection (*Verbindung*) of two measures, e.g., the specific gravity (*Schwere*) of a body as a ratio of its mass to its volume. This ratio is an exponent which in its turn is a measure of a certain body. Then that whole, which is made up of ratios of specified measures becomes a series (*Reihe*) of different ratios of measures. This series is, therefore, characterized by various exponents, and it is here that an independent thing becomes related to other independent things. It represents and displays (*darstellen*) what it is in a series of various exponents, as a series of specific relations to others. Finally, the whole of measure becomes, according to Hegel, an ordered series of ratios of an object to other objects. Each of them is characterized by a specific gravity, and they are related on the basis of the so-called elective affinity (*Wahlverwandschaft*).[7]

What can be learned from Hegel's approach to the categorial cluster "Measure"? Three aspects have to be emphasized here. Firstly, Hegel in this cluster explicitly distinguishes between the external and the immanent measure. Already in its very beginning he claims the following (1923, T. 1, 341; 1969, 331).

The development of measure which . . . starts from the immediate, external measure . . . should . . . indicate the connection between this determination of measure and the *qualities* of natural object, at least in general . . .

Then, when dealing with the specific quantum, the rule, and the standard, he makes this claim about the conventionally chosen unit (1923, T. 1, 343; 1969, 334):

Such a unit, . . . insofar as it is also used as a standard for other things, is in regard to them only an external measure, not their original measure.

With respect to a standard derived from such a unit he, therefore, claims the following (1923, T. 1, 344; 1969, 334):

It is therefore foolish to speak of a *natural* standard of things. Moreover, a universal standard ought only to serve for *external* comparison; in this most superficial sense in which it is taken as a *universal* measure it is a matter of complete indifference what is used for this purpose. It ought not to be a groundmeasure (*Grundmass*) in the sense that the natural measures of particular things could be represented on it and from which, by means of a rule, they could be grasped as specifications of a universal measure, i.e. of the measure of their universal body. Without this meaning, however, an absolute measure is interesting and significantly only as a *common element* (*ein gemeinschaftliches*), and as such it is not *in itself* but is a universal (*Allgemeines*) by agreement.

Secondly, the whole cluster "Measure" from the book *Being* of the *Science of Logic* pertains *exclusively* to the category of the external measure. Neither the category of the immanent measure,[8] nor that of the manifestation of the immanent measure, is treated and explained in this cluster. Even if, according to Hegel, mechanics and chemistry as sciences are connected to different categories, still their understanding of measure belongs to one and the same categorial cluster, namely, *external measure*. This is evident from what Hegel claims initially for the determination of acceleration as it would be applied in Galileo's law of fall and for Kepler's laws (especially for $s^3/t^2 = const.$) in classical mechanics (1923, T. 1, 353–354; 1969, 343)

In regard to the absolute relations of measure, it is well to bear in mind that the *mathematics of nature*, if it is to be worthy of the name of science, must be essentially the science of measures . . . It is a great service, to ascertain the empirical numbers of nature, e.g., the distances of the planets from one another; but it is an infinitely greater service when the empirical quanta are made to disappear and they are raised into a *universal form* of determinations of quantity so that they become moments of a *law* or of measure-immortal services which *Galileo* for the descent of falling bodies and *Kepler* for the motion of the celestial bodies have achieved. The laws they discovered they *proved* in such a way that they have shown the whole range of the particulars of observation to correspond to them. But yet a still higher *proof* is required for these laws; namely nothing else than that their quantitative relations be known from the qualities . . . that are correlated. Of this kind of proof there is still no trace in the said mathematical principles of natural philosophy . . .

This corresponds to the view expressed in the addendum to paragraph 270 of the *Encyclopaedia* quoted above. To determine the force of gravity by which the earth acts on the moon, one has to transform Galileo's law of the fall of a body on earth to the case of the earth-moon relation, while simultaneously taking into account the empirical data about the distance of the moon from the earth (sixty times the diameter of the earth). Such an empirical moment, which enters a theory and represents its central component, is exactly that which, according to Hegel, places the understanding of measure of both mechanics and chemistry into the categorial cluster of external measure. He states this as follows (1923, T. 1, 378; 1969, 365).

The task would be to recognize the exponents of the ratios of the *series of specific gravities* as a *system* from a rule which would specify a merely arithmetical plurality (*Vielheit*) into a series of harmonic nods. The same demand would apply to our knowledge of the series of chemical affinities already mentioned. But science still has a long way to move ahead to accomplish this task, as far ahead as grasping the number of the relative distances of the planets of the solar system in one system of measure.

Thirdly, Hegel has no *positive* philosophical knowledge about the very nature of immanent measure; he frames it only *negatively*: as something different from the external measure. Even if he distinguishes between the external, empirical measure and the internal, immanent measure, and even if he uses terms synonymous with the latter, like "original measure" (1969, 334; 1923, T. 1, 343) or "groundmeasure" (1923, T. 1, 344; 1969, 334), nowhere in this cluster does he define any of them. That Hegel *positively* knows only the meaning of the term "external measure" and that even in the chapter "Real Measure" he deals only with the external, empirical measure are evident from his reflection on price and value. He claims in the *System der Sittlichkeit* (1913, 441; 1979, 121) that

[t]he pure infinity of Law, its inseparability, reflected on the thing, the particular itself, is its equality with others. And the abstraction of this equality of one thing with others, the concrete unity and Law is the *value*; or rather the value itself is the equality as abstraction, the ideal measure. But the really discovered, empirical measure, is the *price*.

In this case, on the one hand, Hegel correctly distinguishes between what one can discover by abstraction – namely, value – and the empirical measure, the price. But, on the other hand, he claims that what can really be discovered is only the empirical measure. Even if he speaks, with respect to value, about the "ideal measure," the latter should be value itself, and not in what value originates – labor. This approach by Hegel is surprising, because, as Lukács proves in (1976), Hegel in his pre-*Phenomenology* period explicitly draws upon the economical works of Adam Smith. Even if the latter determined the quantitative aspect of value of a commodity by the quantity of labor, i.e., by the *quantitative aspect of its immanent measure*, still Hegel did not connect the sense of the term "immanent measure" with the category of substance. This is, as I will show in subchapter 11.3, one of the effects of the construction of the *Science of Logic* on an idealistic basis, one chosen by Hegel consciously and purportedly, the result of the penetration of its idealistic framework into its very categorial structures.

6.4 IMMANENT MEASURE, MANIFESTATION OF IMMANENT MEASURE AND THE SCALE

Let me now analyze Marx's approach to the category of measure. As shown already in subchapter 4.2, he, like Newton, begins in Section 1 of the first chapter of the first volume of *Capital* by determining the natural kind. The "inhabitants" of the space where his thought-operations start are commodities. The phenomenon which characterizes them is that they interact, are exchanged in certain proportions (one coat for x bleaching, y silk, z gold, etc.). To explain Marx's approach here, let me take the case when one coat is exchanged for, say, four grams of gold. One piece (quantum) of something – here gold – is taken and conventionally set as a unit – *one*; its quality being expressed as *[gram]*, *[g]*. A standard can then be defined as a multiple of this quantity (kilogram, ton, etc.). The magnitude of mass is by itself a measure; it holds $m = \{m\}\cdot[m]$. Now such a type of measure is used by Marx to determine and find out what is the (social) substance of the one coat. It is a certain amount of labor, say, *20* "ones," and this gets its phenomenal expression, in the form of an appearance, four grams. The definition in the form of an equation is here $\{4\}\cdot[g] =: k\cdot\{20\}\cdot[labor]$, where k is coefficient so that $k = \{1/5\}\cdot[g]/[labor]$.

But here the first difference between Marx's and Newton's approach to the cause-ground of the effects under investigation enters in; its importance will become apparent below. For Newton the quantitative determination of the cause is just a matter of fact, e.g., in our example from subchapter 6.1, the fact that the quantity of force is twenty units. But for Marx its quantitative determination has its origin in the quantitative moment, or aspect of the cause-ground-essence-producing process. Marx here, with respect to the categories "cause-ground-essence" brings in Hegel's term "immanent measure." But, in contradistinction to Hegel, he gives its sense. Marx determines here, *initially*, "immanent measure" as the *quantity* of the substance from which cause-ground-essence originates. This is readily seen from the following claim (1980b, 109-110; Marx and Engels 1975ff., Vol. 29, 272)

> Labor-time is the living existence (*Dasein*) of labor . . . it is its living existence as a quantitative simultaneously with its immanent measure . . . The labor-time objectified in the use-values of commodities is both the substance, which makes of them exchange-values, and, therefore, also commodities, and it measures their determinate quantity of value (*Werthgrösse*).

Now my claim is here is that Marx, when starting his treatment on commodity in Section 1 in the first chapter of *Capital*'s first volume proceeds – even if with the above given difference – according to categorial patterns of thinking similar to Newton's when Newton moves from Definitions I and II to Definitions III through VIII or when he moves in Book III from celestial phenomena towards gravity. Both take objects of a certain kind, analyze their apparent determinations which are generated by their mutual interactions, and then determine the ground-cause of these determinations. Both take the quantity of certain effects and by means of them determine the quantity of their cause. They start from the *external measure* and proceed in thought by means of it towards the ground-cause-essence. Both Newton's and Marx's movements toward the ground-cause-essence are framed by the knowledge that objects of one kind (commodities, mass-bodies) in their mutual interactions acquire the same apparent properties (exchange-value, acceleration, change of motion); so the cognitive imperative that guides their thinking is here: *same effect, same cause*. This was explicitly stated by Newton as the second rule in the very beginning of the third book of the *Principia*: "to the same natural effects we must, as far as possible, assign the same causes" (1946, 398). This enables him in this book to "derive from the celestial phenomena the forces of gravity with which bodies tend to the sun and the several planets" (1946, XVIII).

The fact that Marx in Section 1 brought in the quantitative aspect of the immanent measure as the quantity of the cause-producing substance enables him to proceed to Section 2 where his thinking behaves according to categorial patterns which are *completely missing* in the *Principia*. In Section 1 Marx proceeds from the appearances (exchange-values) by means of the apparent (external) measure to the ground-cause-essence (value), and then by means of the category of the immanent, internal measure determines its quantity by a certain quantity of labor-time. From the equation *{4}·[g] = :k·{20}·[labor]*, given above, Marx, in Section 1, explains the meaning of *{20}*. In Section 2 he moves to the second aspect, the moment of the immanent measure, to an analysis of the *quality* of the cause-ground essence-producing substance, to the analysis of *[labor]*. What makes Marx's economical works so interesting for contemporary philosophy of science as well as epistemology is the fact that he employs in them *categorial patterns guiding the theoretical analyses of the quality of the very essence-cause-ground*. This also makes them, from the point of view of categories of thinking, superior to those which guide the construction of the *Principia*. In the latter the knowledge of the quality of the essence-cause-ground is given by the empirical knowledge of the quality of the appearances. Newton, therefore, claims that "the qualities of bodies are only known to us by experiment" (1946, 398). For Marx a theoretical qualitative analysis of the essence of objects of a certain kind involves an analysis of its internal differentiation, its contradiction. Because Marx states his views on what is an internal contradiction in the context of his *Critique of Hegel's Philosophy of Right*, I will analyze in detail his views on the internally differentiated essence in chapter eleven by comparing them with Hegel's approach to the category of essence, substance, and contradiction.

Marx views the qualitative aspect of the magnitude of value as internally differentiated (1983, 29) because "the same labor is differently, and even oppositely, determined," (1983, 26) namely, as concrete and abstract labor. Commodity, which comes into being from such an internally differentiated substance "is, therefore, an immediate contradiction"; it has in it an "immanent contradiction" (1983, 53). Because, according to him, the very cause-ground-essence of the phenomena it produces is internally differentiated, it is an active, acting ground.[9] What makes Marx's approach so interesting is the fact that after proceeding from a consideration of the effects as appearances to a discussion of their ground, he is targeting the very ground; *he turns it into the subject of an independent theoretical qualitative inquiry*. His aim is to study not only the ground as something behind the phenomena it generates – the *relative ground* – but to discover also the very ground, what it is in itself, independently from all phenomena (appearances and manifestations) – the *absolute ground* of objects of a certain kind.

Marx's approach to the category of essence by means of the category of the immanent measure solves Ellis's problem given under number seven in subchapter 6.1. Because the immanent measure is the basis (*Grundlage*, in Hegel's words) from which the quantitative and qualitative determinations of the very cause-ground-essence originate, the former determines the latter in an *absolute*, *universal*, and *necessary* way. Marx claims that "the quantity of value of the commodity expresses . . . a necessary relation, which is inherent to its process of production, to the social labor-time"[10] (1983, 64; 1987b, 128). Only where certain essential conditions exist, there exists a certain quantitatively and qualitatively determined substance from which the quantity and quality of the ground-cause-essence is by necessity produced. The ground as generated from the essence-producing substance is the absolute ground. Therefore, the ordering procedure, which already is based on the knowledge of the immanent measure of the ground, represents a type of knowledge of the quantitative determinations of the ground which is necessary, absolutely, and universally related to the magnitude of the ground. In such a type of knowledge there is no need for various external measures. Ellis was right in his contention that the existence of a universal ordering procedure for a magnitude (in his words "quantity") characterizing certain objects under investigation is somehow related to their essence.

There are at least two differences worth mentioning between scientific knowledge which remains at the level of the category of the formal ground and external measure and scientific knowledge which operates already at the level of the category of real ground and immanent measure. Firstly, in the former, the name used to refer to the ground can be viewed as a useful conventional tag or abbreviation for certain phenomena; as an "economical bridge-principle" by means of which the special sciences can, by starting from certain phenomena, arrive at certain other phenomena. For this reason, there is always a temptation to dispose of that name as useless metaphysical ballast. As I will show in the next subchapter, this was in fact done by Ernst Mach in (1960). Contrary to this situation, in scientific knowledge which operates already at the level of the knowledge of the real ground and immanent measure, the ground is understood as something really existing, having its own "life"-process, and the ground is understood as being produced under conditions which are different from the conditions under which the various phenomena are produced. In chapter five the former were labeled inherent conditions, while the latter were labeled in chapter three as modification conditions. At this level of scientific cognition the name of the ground cannot be dismissed as merely a useful tag referring to nothing real and which can, therefore, be disposed of at will.

Secondly, the recognition of the ground as a *formal* ground and, alternately, recognition of the ground as a *real* ground, represent two different types of foundations upon which one can explain or from which one can derive the phenomena as manifestations. As shown already in subchapter 3.5, the knowledge of the ground as formal represents an *incomplete explanatory basis* for the derivation of the phenomena-effects as manifestations of the ground-cause. The problem is that the passage from the knowledge of the appearances to the recognition of the manifestations – by means of the recognition of the cause, is here mediated by a tautological, formal reversal of the definition $f_k(H){:} = \mathrm{E}^{(k)}$ into the "definition" $E^{(k)}{:} = f_k(H)$. That is what Hegel labels as the tautological "retreat into the ground and the emergence from it" (1969, 462; 1923, T. 2, 83), and in that in this case "one finds oneself in a kind of witches' circle" (1969, 461; 1923, T. 2, 82).

The situation is different in the case when the explanation starts from the knowledge of the ground as a real one, i.e., from the knowledge of the internal differentiation of the qualitative aspect of the immanent measure together with the knowledge of its quantitative aspect, its moment. This is the case of Marx's derivation of the concept of exchange-value and money from that of value and concrete/abstract labor. In his view the internal differentiation of the substance of value, its double determination as both concrete and abstract labor, its internal "contradiction must develop itself" (Marx 1983, 53). This contradiction "does not come to a standstill (*ruht und rastet nicht*) until it has shaped itself into the redoubling (*Verdoppelung*) of commodity into commodity and money" (Marx 1983, 53). According to Marx the internal contradiction in an object by necessity develops into an external processual relation of objects, where each, with respect to the other, becomes an extreme, a carrier of one of the poles of the internal contradiction. One of the commodities – gold

– in its processual relation to other commodities becomes the sole, 'pure' representation of value devoid of any use-value; it becomes money. The value of commodities has a quantitative and qualitative moment. Their quality – [labor] – and quantity – {labor} – acquire their expression in the quantitative and qualitative moment of the value of gold. Gold as a value, i.e., as a result of labor, becomes the measure of value. More generally, the measure of the ground becomes in the mutual interaction of entities with the same ground, the manifestation of their common ground, of their immanent measure in their mutual processual relations. Otherwise expressed, the measure of the ground is just the manifestation of the immanent measure of the ground at the level of its effects. The fact that one entity, with respect to another, becomes the expression of the immanent measure of their common ground is only one side of their mutual relations. The second side is that the other entity represents and manifests its own ground in the first one. According to Marx these two processes are intertwined, so that (1980b, 141; Marx and Engels 1975ff., Vol. 29, 305)

> [b]y the same process . . . through which commodities represent their values as gold prices, so do they represent the gold as the measure of value and, therefore, as money.

The thought-derivation of the forms of manifestation of the ground on the basis of the knowledge of the immanent measure of the ground is significant in at least three ways. Firstly, one explains here why by necessity there exist certain phenomena. Scientific cognition initially started from the fact that there are certain phenomena; it presupposed their very existence. By grasping the quality of the ground, its internal differentiation, as well as its quantity, scientific cognition acquires knowledge explaining why these phenomena exist at all, why they are produced. So, for example, in Marx's derivation "[m]oney is the necessary form of manifestation of the immanent measure of commodities, labor-time" (1983, 59; 1987b, 121). This type of knowledge then, together with the knowledge of the process of reproduction of the very ground expressed by means of the inherent organic type of a scientific law, yields the knowledge of why certain types of phenomena are continuously reproduced, why they regularly appear. Thus, there is a difference in epistemic value between the knowledge of the phenomena-effects gained through investigation before one grasps their ground-cause and the knowledge of the phenomena-effects derived in thought from the quantity and quality of their ground-cause. One is dealing here with a growth of knowledge. That growth of epistemic value and of knowledge justify labeling the former phenomena as appearances and the latter as manifestations, a process already described in Chapter 3. The cognitive imperative when proceeding from the effects as appearances to their cause was, as claimed above: *same effect, same cause*. Only after the discovery of what is the cause in itself can one explain why this cause generates certain effects, i.e., only in this way can one arrive at the cognitive norm: *same cause, same effect*. The latter is here of course stripped of its Humean meaning and can be restated as follows: *The immanent measure must manifest itself*. In chapter eleven I will compare this with Hegel's statement "The Essence must manifest itself" (1923, T. 2, 101; 1969, 479).

Secondly, the thought-derivation of the manifestations of the immanent measure at the level of the effects enables us to gain a more advanced knowledge with respect to the external measure and the standard. At the level of knowledge of the measure as an external measure, knowledge is based on a conventional choice of a unit of measurement and on the expression of the quality and quantity of the ground-cause solely on the basis of the knowledge of the quantity and quality of its effects, e.g., for the second dynamic law it holds that $\{m\}\cdot\{a\}\cdot[m]\cdot[a] = :\{f\}\cdot[f]$. But at this level of knowledge it remains a complete mystery why certain effects (say, covered distance, change of velocity, or motion in a certain time) can be accepted at all as the measure of their own causes. Only at the level of scientific cognition based on the knowledge of the immanent measure of the ground and of its manifestations at the level of its effects can this mystery find its rational solution. Entities with this ground interact because the quality of the substance of the ground is internally differentiated. And because in each of them their common substance is quantitatively different, they interact in certain quantitative proportions. In turn, because they interact, certain properties are produced in some of these objects which are the effects of their interaction and, therefore, also of this substance. And because they enter into the process of interaction in certain quantitative proportions, these effects also have certain quantitative determinations. Only because of this, at the level of cognition where there is as yet no knowledge about the very ground – its quality and quantity, is it possible to take certain magnitudes (exchange-value, change of velocity, change of motion, covered distance) and use their quantitative and qualitative determinations, i.e., use their measure to determine the measure of their cause.

Scientists and as well as laymen initially have before them an established maze of mutual interactions, a flux of phenomena (in the words of C. G. Hempel), the cause, the mediating links with their cause, and the genesis of the maze which they have not recognized yet. They, therefore, have to take the results of these interactions, the effects of their common ground, and use them as the starting points of cognition, whose aim is to penetrate behind these phenomena.

Thirdly, through thought-derivation of the forms of manifestation from the knowledge of the immanent measure of the ground, there occurs a fundamental shift in the "space" where scientific thinking takes place; cognition halted at the level of the knowledge of the external measure cannot reach this space. The latter starts with entities of a certain kind (commodities, objects with mass), which by their mutual interactions create that kind as a functioning whole and acquire also certain phenomenal properties-effects (exchange-value, change of motion). But by discovering the immanent measure of the ground it becomes possible both to understand what unifies certain entities into one kind – to understand their common acting ground – and, by explaining why they necessarily interact, to understand why they set up the kind as a functioning, acting whole. Such a shift is exemplified by Marx's move from section 1 to section 3 of the first chapter in *Capital*'s first volume. He starts from commodities and their exchange-values as given in the process of their exchange. Then, by analyzing the very ground of commodities, their value and substance, he gains the ability to derive exchange-values as forms of value, and to explain why the circulation process (commodity → money → commodity) takes place at all.

The development and growth of scientific knowledge, reconstructed in chapter five as the movement of cognition from the pure idealizational type of a scientific law, $L^{(k)}$, to the inherent idealizational type of a scientific law, $L_{ii}^{(k)}$, is here reconstructed as the transformation and reinterpretation of the knowledge of the external measure at the level of the effects as appearances into the knowledge of the manifestation of the immanent measure at the level of effects as manifestations. Because, starting from chapter three, I distinguished two types of knowledge of the phenomena-effects – appearances and manifestations – I arrive here at a differentiation of two forms of knowledge of the measure of the phenomena-effects: the apparent (external) measure of the ground as given by the appearances and the measure of the manifestations, i.e., the immanent measure of the ground as given at the level of its manifestations.

By distinguishing between the knowledge of the effects as appearances of a ground and the knowledge of the effects as the manifestations of this ground, and by simultaneously distinguishing between their respective measures, it becomes possible to find a solution to the two problems stated by Ellis and given under numbers one and five in subchapter 6.1. When physics proceeds from the effects of forces to the very force, the latter's existence is entailed in knowledge by the existence of the former, and the measure of the former is used to determine the measure of the latter. So, as given above, it holds $\{m\}\cdot\{a\}\cdot[m]\cdot[a] = :\{f\}\cdot[f]$, the measure of the magnitude *ma* is here as yet only the apparent measure for the magnitude *f*. If classical mechanics would have the knowledge of what *f* is in itself – knowledge about the immanent measure of absolute force – then it would be possible to derive the existence of its effects from this knowledge. In this case the existence of the effects would be in our knowledge entailed by the existence of forces and the measure of the magnitude *ma* would be already derived here as the manifestation of the immanent measure of the absolute force. Because Ellis does not connect in his approach the category of essence with that of the immanent measure, he does not distinguish between the apparent measure and the manifestation of the immanent measure as two types of knowledge of measure related to phenomena-effects.

To make this point more understandable let me start from the supposition – a rather speculative one – that physics already knows the substance, *S*, of the absolute force; that is, that a certain physical theory already contains the equation $\{f\}\cdot[f]: = \{S_1\}\cdot[S]$. Suppose also that it would also know how to derive the magnitude of mass as it originates in this substance; that is, it would formulate the equation $\{m\}\cdot[m]: = \{S_2\}\cdot[S]$. Then physics could derive the magnitude of acceleration as $\{a\}\cdot[a]: = \{f\}\cdot[S]/\{m\}\cdot[S]$; that is, as a pure quantity given by the ratio $\{S_1\}/\{S_2\}$. If physics would also be able to derive the magnitude of time as $\{t\}\cdot[t]: = \{S_3\}\cdot[S]$, then it could arrive at the equation $\{v\}\cdot[v]: = \{S_1\}\cdot\{S_3\}\cdot[S]/\{S_2\}$ for the magnitude of velocity and finally at $\{l\}\cdot[l]: = \{S_1\}\cdot\{S_3\}^2\cdot[S]^2/\{S_2\}$ for the magnitude of distance. What is readily seen here is that the magnitudes *a*, *m*, *v*, *t* and *l* would be in such a case only quantitative degrees in the manifestation of one and the same substance *S*. Mechanics would in this case be built, from the point of view of quality, by means of the knowledge of just *one* quality, namely *[S]*. This is really the case in Marx's *Capital*. As shown already in chapter five above, in the manuscript of the third volume of *Capital*, the production price is only a quantitative modification of value. Similarly, the various forms of profit are only quantitative "portions" of the magnitude of surplus-value. According to Marx "the surplus-value of individual capital in each particular sphere of production [is] the measure of the absolute quantity (*Grösse*) of the profit," (1980d, 1626; Marx and Engels 1975ff., Vol. 33, 98-99). And because the particular rates of profit "serve as the substance, supposition of the *general rate of profit*," (1980d, 1626, Marx and Engels 1975ff., Vol. 33, 97) for *Capital* as a whole it holds that "the *total surplus-value*, which the total capital . . . produces, [is] the absolute *measure of the total profit of the total capit*al, whereby profit should be understood to include all forms of surplus-value such as rent, interest, etc." (1980d, 1627; Marx and Engels 1975ff., Vol. 33, 99). Keeping in mind that Marx in the first volume of *Capital* determines the substance of value and surplus-value as labor and surplus-labor expressed in units of time, and keeping also in mind that in its second volume, the unit in which the process of the circulation of capital is expressed is also time,[11] then I arrive at the conclusion that the various forms of manifestation, as they theoretically derived in *Capital*, are all expressed on the basis of knowledge of one and the same substance – labor.

These manifestations are the modifications of the quantitative determination of the substance of the ground. A scientific theory which, like *Capital*, succeeds in explaining all phenomena-effects-manifestations $E^{(k)}$, $E^{(k-1)}$, . . ., $E^{(k-j)}$ of a certain kind as only quantitatively differently determined moments, degrees of one ground-cause-essence H, is built in a *substantio-monistic way*. The crucial indicator that a theory is constructed in such a way is that even if its various subtheories derived by the method of gradual concretization ($L^{(k)}{}_{ii} -_{gc} \dashv L^{(k-1)}{}_{ii} \ldots -_{gc} \dashv L^{(k-j)}{}_{ii}$) express the knowledge of different quantities $\{x_1\}$, $\{x_2\}$, . . ., $\{x_n\}$, they all still express the knowledge of one quality $[x]$, where the latter stands for the quality of the ground–essence.[12]

The answer to Ellis's question given under number four in subchapter 6.1 can therefore be given as follows: Because (on the supposition that physics already knows the immanent measure of absolute force) the magnitude *ma* functions both as the apparent measure to determine the measure of the force and as the manifestation of the immanent measure of the force, there are two types of relations between *ma* and *f*. In the case when cognition is at the level of knowledge of the measure as external measure, this relation has the character of a non-necessary *a posteriori* process used to determine the measure of the force. But what lies behind this seemingly accidental process of identification is the existence of a real (*de re*) necessary production of the change of motion in the process of mutual force – interaction between objects. After the immanent measure of force has been recognized, the magnitude *ma* would be by necessity entailed from the knowledge of the immanent measure. Its derivation would explain why force by necessity manifests itself as acceleration.

Now it is possible to compare the categories of Marx's thought in his economical writings with those of Newton's in the *Principia*. Newton, like Marx, understands the requirement to proceed in cognition from the appearances – the interaction of objects of a certain kind – to their ground, and then from this ground to its various manifestations. Newton, like Marx, is well aware of the requirement to grasp the absolute ground (force), but unlike Marx, he does not provide the knowledge of its immanent measure. Newton discards the notion of the *absolute* force in the *Principia* and remains at the level of the categories of the formal ground and external measure. Marx, on the contrary, explicitly operates with the concept of *absolute* value (1992, 118, 310, 334). He moves from categories of apparent measure and formal ground to those of the real ground and immanent measure in order to arrive finally at the category of the manifestation of the immanent measure. These are the *three faces of measure*. Marx as a scientist, in comparison with Newton as a scientist, operates in the framework of more profound, richer, and more advanced categorial structures of thinking. This is yet another proof – the third in addition to those two given in chapter five – that Marx is not the Galileo of the social sciences, as is claimed by L. Nowak in (1980). In the conclusion of this book I will prove that Marx is not the Galileo of the social sciences for yet another reason, namely because he analyzes and explains social phenomena, while Galileo and Newton explain natural phenomena, i.e., because the former is a social scientist, while the latter are natural scientists.

6.5 ERNST MACH

By distinguishing the three types of knowledge of measure – the external, the immanent and the manifestation of the latter – it is possible to analyze Mach's attempt to dispose of the concept of force. Mach attempted to purge mechanics from the concept of force as something causal and replace it by an understanding of a kind that would allow it to be derived as a purely mathematical expression in the form $F: = ma$. For the first time he explained his approach in his 1868 article; I explain it by drawing upon *The Science of Mechanics* (1960). He puts forward the following statement: "All those bodies are bodies of equal mass, which mutually acting on each other produce in each other equal and opposite acceleration" (1960, 266). He generalizes this case and shows that if bodies A and B receive respectively, as the result of their mutual action, the acceleration with the absolute value of a_A and a_B, then body B has a_A/a_B times the mass of the body A. Then "[i]f we take A as our unit we assign to that body the mass m which imparts to A m-times the acceleration that A in reaction imparts to it" (1960, 266). Finally, he claims that "[m]y definition is the outcome of an endeavor to establish the interdependence of phenomena and to remove all metaphysical obscurity" (1960, 267).

To understand Mach's generalization let me analyze[13] how Newton deals with the relation of two bodies, designated here as A and B, after he has dealt with the accelerative, motive and absolute force. By choosing the former mass m_A as a unit it is possible to express the mass of the body B as a multiple or fraction of m_A. Body A is here in such a position that B can express its mass as equivalent or as a degree of equivalence with A. So A is here, with respect to B, in the position of an equivalent form. Body B is here in another position. What its mass is is expressed with respect to (i.e., relatively to) A; it is in a relative form. This can hold also the other way round; one could take the mass of B and set it as a unit. So not only does A provide its body to express something belonging to B, but also B's body can be used to express something belonging to A. This something is, according to Newton, force f. In the former case force f_B tied to the body B uses the body A to manifest itself as a_A. In the latter case f_A tied to the body A uses the body B to manifest itself as a_B. In the latter case A is in the position of the relative form while body B is in the position of the equivalent form; in the former case it is the other way round. The basis (Hegel's *Grundlage*) of the fact that the mutual relation of A and B is that of the relation of an equivalent and relative form or[14] *vice versa* is the existence of something different from acceleration and mass – it is force. It exists as a common ground in both bodies. What is common to both of them is one quality $[f]$ given in each of them in quantities $\{f_A\}$ and $\{f_B\}$, so that $f_A = f_B$ or $\{f_A\}\cdot[f] = \{f_B\}\cdot[f]$ holds. On both sides of this equation we have the same quality and the same quantity. This equation is, in my view, present in Newton's third law: "the mutual actions of two bodies upon each other are always equal, and directed to contrary parts" (1946, 13; 1687, [13]). His claim regarding this equality is based on what he labels in Definition VI as "absolute force." This is readily seen from the commentary on the third law (1946, 14; 1687 [13])

If a body impinge upon another, and by its force change the motion of the other, that body also . . . will undergo an equal change, in its own motion, towards the contrary part.

So this law in fact claims that "the mutual force-actions of two bodies are always equal and directed to contrary terms." What has to be emphasized is that only if there is a really existing ground of a certain kind on the basis of which objects enter into certain mutual ratios, can special science initially determine the quantity of certain apparent properties, the effects produced on an object, and where these effects are at the level of manifestations, therefore, in the position of an equivalent form. Hence, by conventionally setting the quantity of these phenomenal properties as one, we may determine their quantitative determinations in other objects. What has to be also emphasized is that because Newton in the third law and its commentary starts from the relation $f_A = f_B$, he can arrive at the claim that $dp_A/dt = dp_B/dt$ holds (i.e., that $m_A a_A = m_B a_B$ holds), and therefore finally also claim that "the changes of the velocities made towards contrary parts are inversely proportional to the bodies" (1946, 14; 1687, [13]). Mach accepts this result of Newton's process of thought, i.e., that $a_A/a_B = m_B/m_A$, but disposes of both $[f]$ and $f_A = f_B$. What is then left over is only a statement concerning the relation of phenomena-effects, as it is given in the last of Mach's quotations presented above: a_A to a_B. But one has to bear in mind that this is only a *post festum* disposal of Newton's propositions regarding force. This kind of rejection can be accomplished only afterwards and only because Newton had already accomplished, prior to Mach's rejection, the following thought-movement. He starts from the apparent forms (change of motion and speed in a certain time) and, by means of them, derives the concept of the accelerative and motive force. He then introduces the concept of absolute force and arrives, via $f_A = f_B$, at $a_A/a_B = m_B/m_A$.

But this thought-movement bears in itself a fundamental problem. Newton, correctly (at least in my view) starts from the fact that bodies mutually interact, and through these interactions acquire certain phenomenal properties which are quantitatively determined. Then he tries to grasp the ground of these interactions by passing over to the concept of force. Finally, Newton wants to arrive at an explanation of these interactions by means of his three laws (Lex I, II, III); i.e., to explain the phenomena as manifestations of force. In the first law the change of the state of a body is *reasoned by the concept of force* acting on it. In the second law the change of motion in time is also reasoned by the concept of force. Finally, in the third law the action and reaction of bodies in their mutual interaction is reasoned by the concept of force as well. But because Newton, as shown above, does not reason the absolute force, in his *Principia*, it remains a complete mystery why bodies mutually interact. Because he does not know what $[f]$ is, he has to claim in the second law, that "[t]he change of motion is proportional to the motive force impressed" (1687, [12]; 1946, 13). But because the motive force is defined in definition VIII as proportional to the change of motion in time, Newton's movement from the *appearance* given in this definition to the *manifestation* stated in the second law ends up in the trivial claim that the change of motion of a body in a certain time is proportional to the change of its motion in this time. For the same reason, namely, the absence of knowledge of what $[f]$ is, Newton can neither derive the relation $\{f_A\}\cdot[f] = \{f_B\}\cdot[f]$ as it is given in the third law.

The issue of interactions enables me to look at Mach's phenomenological, antimetaphysical attitude from yet another point of view. He presupposes that bodies interact; only then can he apply his phenomenological, antimetaphysical attitude. But the fact that bodies interact was already realized by Kepler for celestial objects and by Galileo and Huyghens for terrestrial objects. Newton, starting from the knowledge produced by them – that bodies interact in certain proportions and due to these interactions acquire certain properties in a certain quantity – pushed farther, impelled by the ambition, firstly, to discover the very ground of these interactions and of these proportions and quantities and then, secondly, to derive the former from the latter. Mach's ambition, contrary to Newton's, is just to remain at the level of knowledge which fixes the fact that bodies interact in certain proportions. From the point of view of scientific knowledge it represents a retreat from, and a regression with respect to, the level of knowledge attained in the *Principia*.

6.6 COEFFICIENTS AND UNIVERSAL CONSTANTS IN PHYSICS

By distinguishing three types of knowledge of measure it is possible to address problems related to coefficients and constants appearing in the equations of physical theories.

Let me start with the coefficients which one encounters in the law of the expansion of a solid and in the equation for the pendulum $T = 2\pi\sqrt{l/g}$ as given before its derivation from Newtonian mechanics. The general equation for the expansion of a solid is $x_t = x_0(1 + \alpha t)$; t denotes the rise in temperature producing the increase of the magnitude of volume, x, from its initial value $\{x_0\}$ to its final value $\{x_t\}$; α is the appropriate coefficient of expansion varying from body to body. In Huyghen's derivation of the equation for the pendulum, g appeared as a coefficient which, even if known to be fixed in certain locations, was known to be different at various spots on the earth. Huyghens' assistant, J. Richer, noticed in 1671 in French Guinea that in order to acquire here a conformity between the time as measured by his pendulum clock and the motion of the Sun, he had to shorten it from its length in Paris (see on this (Jammer 1961, 73–74) and (Sarlemijn 1993)).

What do the coefficients α and g have in common? They represent an important piece of knowledge from the point of view of knowledge of quantity. Physics, initially, starts in its various theories from the quality of distance, time, mass, etc. By deriving the equations like $x_t = x_0(1 + \alpha t)$ and $T = 2\pi\sqrt{l/g}$ it arrives at the knowledge of certain quantities it knew nothing about before and finds out also that these quantities are varying from body to body and from one location in space to another more distant one. Such knowledge then drives empirical investigations into discovering values for different bodies and locations. But regarding *quality*, nothing new was added to the stock of knowledge. This is readily seen if one takes, e.g., the coefficient g, and analyzes it to find out its quality. It holds $[l]\cdot[t]^{-2} = :[g]$. Classical mechanics starts with the qualities $[m]$, $[l]$, and $[t]$, and after deriving $T = 2\pi\sqrt{l/g}$ still remains at the level of the knowledge of these qualities; nothing new was added to the latter. The same holds for the coefficient α. It holds $[T]^{-1} = :[\alpha]$, where $[T]$ stands for [absolute temperature]

The situation is similar in the case of the universal constants like speed of light, c, the gravitational constant, G, Planck's constant, h, and Boltzmann's constant, k. Even if physics found out that, at least up to this point in time, these constants do not change from body to body or from one point of space to another one, still – if one speaks of quality – they share a profound similarity with the coefficients analyzed above. Let me analyze the gravitational constant, for example. It unifies in itself a quantitative moment, $\{G\} = 6.6735 \pm 0.0031 \times 10^{-11}$, and a qualitative moment, $[G] = [kg]^{-1}\cdot[met]^3\cdot[sec]^{-2}$. Mechanics arrives at G hand-in-hand with the derivation of the law of gravitation, starting from the magnitudes of mass, distance, and time. At the very beginning of this derivation their quantitative determination $\{m\}$, $\{l\}$, and $\{t\}$ is not given. But by deriving the law of gravitation, mechanics ends up with the knowledge which leads to measurements yielding the knowledge of a new quantity, namely $\{G\}$. But, as in the case of the above-analyzed coefficients, one starts with $[m]$, $[l]$, and $[t]$, and when deriving $[G]$, one concludes by merely reasserting their combination, $[m]^{-1}\cdot[l]^3\cdot[t]^{-2}$. The same holds for c, h, and k: $[l]\cdot[t]^{-1} = :[c]$; $[m]^2\cdot[l]^2\cdot[t] = :[h]$; *and* $[m]\cdot[l]^2\cdot[T]^{-1} = :[k]$.

By distinguishing the three faces of measure, it is also possible to understand the twofold function of coefficients and constants appearing in the equations of scientific laws. They are present in these equations by their quantitative moment because they mediate between two sides of an equation which differ from one another quantitatively. Behind this difference is, in my view, the fact that these sides express quantitatively different aspects – metamorphoses, modifications – of one substance. But the presence of the qualitative moment in the coefficients and constants of nature is of a different nature. This moment is present in an equation of a special science only so long as this science has not reached the true turning point in the movement of cognition from comprehending the phenomena as appearances to comprehending the phenomena as manifestations: the knowledge of what is the quality of their ground-cause-essence. As long as a special science has not discovered the immanent measure of the ground of the phenomena it investigates, it has to determine the ground by means of the quantitative and qualitative determinations of its phenomena, i.e., by a measure which is external with respect to this ground. But in order to formulate an equation in such a situation, i.e., that on both sides there appear the same quality, one has to adjoin a constant to the expression on one side of it so that together they yield a quality which is the same as the quality given on the other side of the equation.

What is the position of coefficients in equations of a scientific theory which is built by the method of explanation by gradual concretization which starts from the inherent idealizational type of scientific law? Primarily, because all its subparts, its subtheories, are derived already from knowledge of the immanent measure, and because they express only degrees in the quantitative modification – metamorphoses of the quantitative determination of *one quality* – these constant are devoid of any quality. Their function is to mediate between two sides in this type of an equation, where one side represents a certain manifestation which differs from the manifestation expressed by the other side quantitatively, but where each of them expresses one and the same quality.

By drawing upon the reconstruction of scientific explanation by gradual concretization based on the inherent idealizational type of scientific law by means of gradual concretization in chapter five, the gradual derivation of manifestations from other manifestations can be symbolized as follows ("h" stands here for a certain function):

$$E^{(k-1)}(S, Cmod_k(S)) := E^{(k)}(S) + h(Cmod_k(S))$$

or, by means of that purely quantitative constant c, as

$$E^{(k-1)}(S, Cmod_k(S)) := cE^{(k)}(S)$$

The quantitative difference between the two sides in this equation, which necessitates the introduction of that constant, is the result of the introduction of the modification condition $Cmod_k(S)$ in the process of the gradual concretization of the inherent idealizational type of a scientific law $L_{ii}^{(k)}$ into that of $L_{ii}^{(k-1)}$. Because both sides of this equation express, as unified in one equation, the ratio between two different quantities, the quantity of c mediates between them and enables them to express their relation and ratio in the form of a valid, meaningful equation.[15]

CHAPTER 7

THE ESSENTIAL FORMALITY

The aim of this chapter is to reconstruct certain aspects of the categorial framework of Marx's second volume of *Capital.* The reconstruction of the inherent types of a scientific law in chapter five showed that the essential conditions have the character of the cause, and the essence-producing process is their effect, while in the process of reproduction they continually exchange these functions mutually. So in the inherent organic and in the developmental type of a scientific law, the causal characteristics of the object under scrutiny are understood as having the character of causal loops, i.e. understood in a non-linear manner. Given this proposition, the reconstruction of the categorial framework of the second volume of *Capital* can have certain implications for the philosophical analysis of those theories in contemporary physics, physical chemistry, and biology which grasp their objects in a non-linear manner. As an attempt at this we apply the results of the analysis of second volume of *Capital* for the reconstruction of the foundations of non-equilibrium non-linear thermodynamics as given in the works of Ilya Prigogine. In chapter nine I will also subject to analysis from this point of view Manfred Eigen's concept of the hypercycle.

7.1 THE LOCAL ESSENCE-PRODUCING PROCESS AND THE PROCESS OF ESSENTIAL FORMALITY

While introducing the notion of the essence-producing process as that of a reproduction process I left unremarked the fact that the process of reproduction involves the input of material from the surrounding area; or, to be more precise, the drawing in and interiorization of material from the surrounding area. The very essence-producing process has from this point of view the character of a substance-subject,[1] which triggers fluxes in its surroundings. The fact that the essence-producing process triggers fluxes signifies that it draws in material, that the latter is not pumped in from the outside by a mechanism independent from the essence-producing process.

The existence of fluxes shows also that our reflections on the essence-producing process in chapter five were based on the supposition that the process of interiorization is not independent from the essence-producing process. Marx viewed the process of production and the process of circulation "as the two great distinctions" (1981a, 418; Marx and Engels 1975ff., Vol. 28, 441), two moments which set up a whole, "a totality of these two processes" (1981a, 506; Marx and Engels 1975ff., Vol. 29, 8). The process of circulation with respect to its unity with the process of production fulfills an essential task (Marx 1981a, 434; Marx and Engels 1975ff., Vol. 28, 459), as its introductory process is its inherent moment (Marx 1988b, 55, 128; Marx and Engels 1975ff., Vol. 34, 465). At the same time, Marx views the process of circulation as a preparation for the triggering of the production process, not as the essence-producing process itself; the essence does not come into being in it. The process of circulation is only a necessary formality, an *essential formality* (Marx 1988b, 129; Marx and Engels 1975ff., Vol. 34, 465).

Another supposition in my explanation of the essence-producing process in chapter five was that the entities entering into the essence-producing process – its presuppositions – were only its own past results, and not the results of other, qualitatively identical essence-producing processes. I also supposed that the results of the essence-producing process, described as reproduction, are not interiorized into other, qualitatively identical, essence-producing processes. Both these suppositions were the results of the *inherent analysis* based on the formulation of inherent types of a scientific law and oriented exclusively towards a reconstruction of qualitatively identical objects. The inherent analysis grasps the essence-producing process as a localized process, whose interactions with essence-producing processes located elsewhere have as yet to be explained. Therefore I will from now on use the term "local essence-producing process."

7.2 REPRODUCTION AND SPACE-TIME STRUCTURE

As I will show in subchapter 7.4, an important point is found in Marx's approach to the relation of the process of essential formality to the essence-producing process. Namely, he views this relation not as a succession of two completely different processes mediated by the essence-producing process, but as a relation where the process of essential formality has a *cyclical* character, i.e., not as . . . $\rightarrow EF_1 \rightarrow EP \rightarrow EF_2 \rightarrow$. . . ($EF_1$, EF_2 stands for the two different processes of essential formality; *EP* stands for the essence-producing process), but as

Figure 1.

Regarding the unity of these processes, the process of essential formality is the starting and final point for the essence-producing process and *vice versa* (1981a, 435; Marx and Engels 1975ff., Vol. 28, 460).

This approach enables Marx to scrutinize the moments given in the very process of essential formality. He can examine, on the one hand, the presuppositions of the essence-producing process, *Pr*, and on the other hand, its results, *Rs*. Their relation is symbolized as *Pr* . . . *EP* . . . *Rs*, where the points indicate that the process of essential formality was interrupted by the essence-producing process. If, as Marx claims, the process of essential formality has a cyclical character, then the results of the essence-producing process should processually change into its presuppositions. According to Marx, this change is mediated by money as the measure, *Ms*, which fulfills the function of a link between the essence-producing process (in its unity with its process of essential formality) with other local essence-producing processes. But Marx, nonetheless, primarily investigates only one local essence-producing process plus its corresponding process of essential formality, and only then explains the unification of local processes into a global one. So, his initial conceptual space in the manuscript of the second volume of *Capital* can be symbolized as follows:

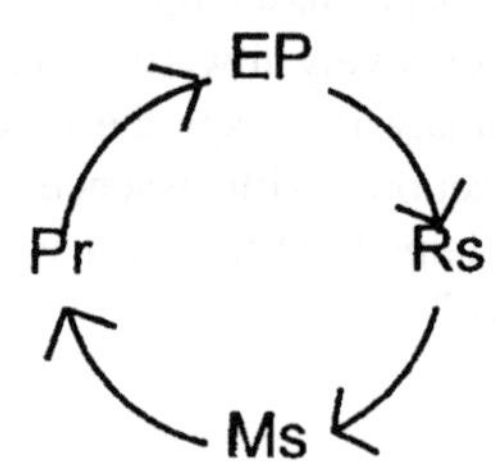

Figure 2.

Pr, *EP*, *Rs*, and *Ms* stand for entities-processes which are moments in the local process; they are four forms of the existence of this process. They mutually change one into another and through their cyclical-processual unity they are themselves "moments of its production – reproduction, as well as new production" (Marx 1981a, 418; Marx and Engels 1975ff., Vol. 28, 440).

Marx's approach to the local process enables him to grasp its space-time determinations. He claims that the local process as a unity of its four forms "is also their difference, a space-like as well as a time-like difference" (1981a, 508; Marx and Engels 1975ff., Vol. 29, 10). Marx's initial approach is based on the assumption that the local process in a certain moment exists only in one of the following four forms:

(i) Pr . . . → EP . . . → Rs → Ms → Pr
(ii) EP . . . → Rs → Ms → Pr . . . → EP
(iii) Rs → Ms → Pr . . . → EP . . . → Rs
(iv) Ms → Pr . . . → EP . . . → Rs → Ms

as well as on the assumption that in each of them the local process can exist only in one of its phases. But then Marx gives up these suppositions, claiming that they are "existing solely for the observer" (1967, Vol. 2, 101; Marx and Engels 1975ff., Vol. 36, 106). In reality the local process exists contemporaneously in all its four forms; they exist *side by side in space* (Marx 1988c, 180), and each of its forms passes through its own process of reproduction (Marx 1967, Vol. 2, 104; Marx and Engels 1975ff., Vol. 36, 109). So the scheme (7.1) changes into

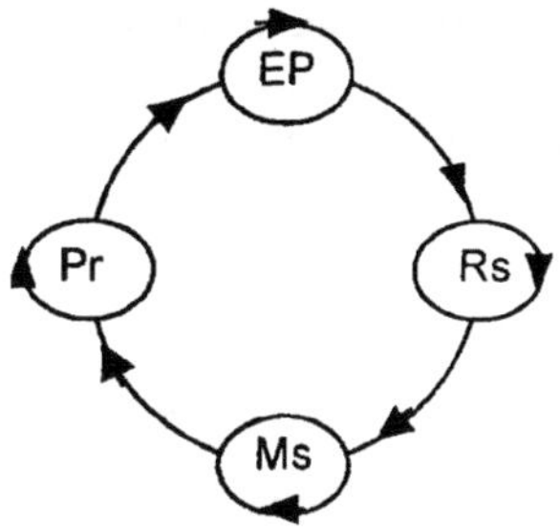

Figure 3.

The real local process of reproduction is a unity of four forms separated in space. On the one hand, the fact that their time-succession exists is the result of the fact that they exist side by side. So the triggering of the essence-producing process from its presuppositions in a certain location is enabled by the processual transformation of the essence-producing process into its results in another location. On the other hand, the fact that there exist four forms side by side – i.e., we are speaking here of their space-determinations – is only the result of the fact that their time sequence exists, a sequence in which one form changes into another one (Marx 1988c, 180). With respect to the forms of the local process, we claim that their space-coexistence is a result of their time-successive existence, and *vice versa*. The time and space structures of the local process mutually generate each other, i.e., the local process by its self-movement generates its own space-time structure.

From the point of view of the space-determinations of the local process, we can distinguish the space of the local essence-producing process as well as the space of its corresponding process of essential formality. An important feature of the space determinations of the local essence-producing process as reconstructed in Marx's *Capital* is that they are a product of this process; *the essence-producing process* is the subject of its own spatial boundaries. Their time-dependence or time-independence is not a result of some external constraints, i.e., of constraints which are qualitatively independent from the essence-producing process, but a direct result of the type of reproduction characterizing the essence-producing process. In the long run, non-simple reproduction leads to a time-dependent boundary, while simple reproduction leads to a time-independent boundary.

Regarding the time determinations of the local process, we can distinguish several types of it in Marx's *Capital*:

a) the time of the essence-producing process, i.e. the time of its duration, the stretch of time that flows between its onset and its end;
b) the cyclical time, i.e. the time given by the span between the onsets of two successive essence-producing processes which do not differ quantitatively;
c) the directional time as the time given by the span between the onsets of two successive essence-producing processes which do differ quantitatively, i.e., the

period demarcating two qualitatively identical points of two successive cycles of non-simple reproduction;

d) the historical time as the time given by the span between the historical becoming and decline of certain essential conditions, between the becoming and decline of a real law (of nature or society).

7.3 THE GLOBAL PROCESS

The global process is understood by Marx primarily as a result of the mutual interactions of the local processes mediated by the measure *Ms*. He scrutinizes these interactions in the various forms in which they exist: return into itself; unidirectional progression; bi-directional progression; circumvention (Marx 1988c, 367). Progression of these types of processes represent a certain time-progression of processes. We have already shown in the previous subchapter that the succession of processes (*das Nacheinander*) creating the local process is possible only if they take place side by side, if they are coexisting (*das Nebeneinander*), and *vice versa*. The same holds also for the global process. It represents a processual time-sequence of forms which take their course "*simultaneously* in their incipient forms, successive forms, and in their last, concluding forms side by side" (Marx 1988c, 368).

Yet one has to bear in mind that Marx views the global process not only as a mere result of the interactions between the local processes. On the one hand, the very local process can take its course only as a part, a moment of the global process which modifies the local process. But, on the other hand, in scientific cognition it is necessary to abstract temporarily from the complex of these processes and from these modifications, and to choose one of them as standing for, representing all the others. So even if the local and the global processes mutually interact, in scientific explanation according to Marx "to depict this complex as such is one thing, another thing is to depict those isolated movements of which it consists" (Marx 1981b, 50). The representational function of the thought-object I have tagged as "local process" is safeguarded by two facts. Firstly, that one operates with this thought-object on the basis of knowledge of essential conditions of the essence-producing process. Everywhere and only there where these conditions are given is it possible to use the concrete thought-object tagged "local process" for a theoretical selection of an object and for its scientific explanation. Secondly, that one temporarily abstracts from all the connections of the local process with the remaining local processes belonging to the same global process.

But Marx's selection of a local process as representing and standing for all the others is possible only by means of a move applied by him consciously in the construction of his theory of capital. Even though in the explanation of the local process Marx abstracts from its connections inside the global process, in fact they are present. The chosen concrete local process is quantitatively codetermined by the interactions with other local processes in the global process. The global process by itself codetermines the local process, which in its turn determines this global process. How will this mutual interaction of the global and of the local influence the explanation of the global? The move applied by Marx consists firstly of the thought-supposition that from the moment of the selection of a concrete local process, the latter has been already quantitatively established, and that from now on it is functioning quantitatively solely on its own ground. This supposition is legitimized by the fact that the inherent analysis is initially oriented only towards the qualitative determinations of the local process. Then it is possible to explain how the sum of the movements of many local processes shapes, generates the global process. This approach was applied by Marx in the first and second volume of *Capital*. In the first he explains the production phase of the "life-process" of one industrial capital. In Part I and II of the second volume he explains the whole "life-process" of one industrial capital. Finally in Part III of this volume he explains the "life-process" of the aggregate social capital given as the sum of the "life-processes" of many industrial capitals. So, the first volume of *Capital* is an exemplification of scientific explanation of the qualitative self-determination of a local process, while the first two parts in its second volume exemplify scientific explanation of the *quantitative* self-determination of this local process. Finally the third part of this volume exemplifies scientific explanation of how local processes determined in such a way generate a global process.

But the explanation of how the global process quantitatively codetermines the local process must in the long run be provided, because the explanation of the very local process and the explanation of the forming of a global process (by the local processes) is done under the supposition that in the end the codetermination of the local by the global will be explained. Otherwise those two explanations would be based on a questionable, problematic foundation. This represents the *second* aspect of Marx's thought-move in his capital theory. I interpret the following statement of Marx in the sense of these two aspects (1967, Vol. 2, 97; Marx and Engels 1975ff., Vol. 36, 103):

> The fact that the social capital is equal to the sum of the individual capitals . . . , and that the aggregate movement of social capital is equal to the algebraic sum of the movements of the individual capitals, does not in any way preclude the possibility that this movement as the movement of a single individual capital, may present other phenomena than the same movement does when considered from the point of view of a part of the aggregate movement of the social capital, . . . , and that the movement simultaneously solves problems the solution of which must be assumed when studying the circuit of a separate, individual capital instead of being the result of such study.

This means that the summary scientific explanation of a global process, exemplified by the explanatory structure of Marx's *Capital* as a whole,[2] has a circular character. It starts intentionally from partially unexplained presuppositions. The whole following sequence of scientific explanations is built on them, and these, and only these, explanations enable the summary scientific explanation to end in the final, complete explanation of its own presuppositions.

7.4 THE FOUNDATIONS OF THE NON-LINEAR NON-EQUILIBRIUM THERMODYNAMICS

Let me now analyze the foundations of non-linear non-equilibrium thermodynamics, an interdisciplinary scientific theory, which was formed in the last thirty or so years,

contradiction by taking into account "the unity of law and of its conditions" (Griese 1977, 127). The point is that (Griese 1977, 128)

> the state and evolution of the cosmos are determined not only by laws of nature themselves, but follow from them together with the initial and boundary conditions which will outline the framework for the operation of simple laws of nature. If those conditions are time-dependent, then, even in the case of time-independent simple laws, it leads to time-dependent, developing systems.

As seen above, the history of a certain object is identical with its trajectory; that is why explaining the concrete history of an object means explaining why the object has undergone such and such a temporarily irreversible succession of states. The alleged succession, however, cannot be explained solely on the basis of a temporarily reversible scientific law. If one is anxious to explain a real becoming of this or that state (say in classical mechanics the concrete values p_2, q_2 in time t_2) on the basis of the preceding state, three options are available. One may proceed from the assumption that the state has arisen under certain initial conditions in a certain time (e.g., from values p_1, q_1 in time t_1). Then, however, the question of their origin arises. The answer that they have emerged out of other initial conditions (e.g., p_0, q_0 in time t_0) cannot be taken seriously. It results in an infinite regression, for we here attempt to ground some initial conditions by means of some other initial conditions. The only foundation for the emergence of certain initial conditions is the fact that they have arisen through the impact of the surroundings upon the object. From the instant of a determination of the initial conditions of the object now moving onward, the object will have been developing along a certain specific trajectory with an irreversible history of its own, whose origination resides in the environmental influence upon the object of investigation.

The second option is this: one may accept that a concrete irreversible history derives not from a certain type of initial (generated from the outside) condition, but from a certain type of boundary (spatial) condition of the object. Now the description of the concrete, irreversible history of the object has been arrived at through a scientific law describing the control of the temporal change of the object's boundary by a law of nature, which should be independent from the law of nature regulating the motion within the system. Those two laws of nature need to be mutually extraneous, hence the trajectory needs to have originated from the action of the surroundings on the object. Physics, then, within the framework of dynamic theories, explains the history of the investigated object only by the description of the initial conditions, or only by the description of the boundary conditions, or else by the concurrent description of both types of conditions.

This type of explanation of an irreversible history of an object is founded on a certain understanding of the concept "law of nature." It is assumed that the law nature which controls the inside of the object operates always and everywhere, in each point of its trajectory-history. It is supposed to have operated before the rise of a certain type of initial condition and to be valid in the process of the surroundings' impact upon the object, an impact that determines the boundary circumscribing the object, as well as the situation afterwards, when this impact changes into another type of impact. The explanation of the irreversibility of the trajectory-history of an object, if built upon such an understanding, cannot be based on a description of its inner motion. It requires in the description "the passage from one system to another," (Wahsner 1977, 146); the passage to the description of the surroundings, which, from the outside, will determine the history-trajectory of the object. It is commonly asserted that (Wahsner 1977, 146)

it is just the situation that usually attends in developmental processes. While a law applies to a certain system, whose surroundings comes forth at the best as a boundary condition, in developmental processes the system and its surroundings are equal, since evolution is such a relationship between the system and its surroundings in whose framework the stimuli are responded to by adaptational changes.

A necessary consequence of the examined approach is also the fact that "the property of time to posses directionality cannot be grounded within the system itself" (Wahsner 1977, 146).

Let me compare the approach to time-direction and history in (Griese 1977; Treder 1977; Wahsner 1977) with the approach to time-direction and history based on scientific inherent types of laws. Even if they apply the idea of unity of a law of nature with its conditions, still the above-mentioned approach does not reach the level of a deeper understanding of this unity. It presupposes that the very laws of nature are independent from space-time conditions and *vice versa*. Contrary to this, in the approach based on the idea of inherent laws, a law of nature is understood not as an unconditioned universality, but as a form of reproduction of the essential conditions, i.e., as an in principle, conditioned form of universality,[3] which is at work everywhere and only there where the respective conditions are at work. This approach also, as I have shown in subchapter 7.3, enables one to introduce the concept of directional time and so to grasp the direction of time *inside* the very object of investigation.

In thermodynamics a central role is played by the second law. It is possible to derive from it the concept of entropy (together with the concept of absolute temperature) by means of which, in turn, it is possible to describe the development inside a completely isolated system towards a state of thermodynamic equilibrium (i.e., towards the state in which all its macroscopic processes cease and its internal organization disappears).

All the hitherto indicated relations of thermodynamics are satisfied only if the system is isolated from the fluxes coming to it from the surroundings (Prigogine and Stengers 1984, 128). As Prigogine puts it, the apparatus of thermodynamics must be enriched by novel relations and concepts in case we aspire to investigate an open system. Then, equilibrium thermodynamics passes on into *non-equilibrium linear thermodynamics*. According to this theory, a system "evolves towards a stationary state characterized by a minimum entropy production compatible with the constraints imposed upon the system" (Prigogine and Stengers 1984, 138). A system which by external factors is constrained from reaching a state of equilibrium will reach a state of minimal inner activity which still can be associated with the fluxes maintaining this activity. So, within the system the situation is similar to situations described in equilibrium thermodynamics; inside the system "the evolution retains the character of the degradation process of 'order → disorder' type" (Glansdorff 1979, 4). If, therefore, the evolution inside a system again leads to the destruction of its internal organization, the following question arises: what are the presumptions preventing its decline? For the sake of discovering these, Prigogine turned to the theory of stability (uncovering of Lyapunov's function) and proved that there is no point in reasoning on the emergence of instabilities except instabilities characterized by chemical processes of the type $X + Y \rightarrow 2X$. Yet this type of reaction is no longer linear. In a non-linear domain it is only plausible to think of the emergence of instabilities.

7.4.3 Non-equilibrium non-linear thermodynamics

Within the whole body of Prigogine's works on non-equilibrium non-linear thermodynamics the focal place is occupied by the so-called Brusselator. It is a thought construct (for its criticism in the perspective of physical chemistry see Gray and Morley-Buchanan 1985; for its defense see Lefever and Nicolis and Borckmans 1988; on its more recent developments see Otwinowski 1990; on its recent application see Ackermann and Wlotzka and McCaskill 1998) involving the following chemical reaction (leaving aside reverse reactions and making the kinetic constants equal to one):

$$A \rightarrow X$$
$$B + X \rightarrow Y + D$$
$$2X + Y \rightarrow 3X$$
$$X \rightarrow E$$

From the surroundings, components *A* and *B* enter into a system, consisting of the components *X* and *Y*, whereas the components *D* and *E* come out from it. By means of the solution of a system of nonlinear differential equations, describing the time change of the concentrations *x*, *y* of the components *X* and *Y* (on the assumption of unchangeability of the concentrations *a*, *b* of the components *A*, *B*; on the assumption of abstraction from diffusion; and on the assumption of a time-independent boundary) it turns out that, for the ratio of the concentrations $b < 1 + a^2$, the system is always in a stable state. However, as soon as *b* reaches a value higher than $1 + a^2$, there emerges an instability and the concentrations *x*, *y* develop in a limit cycle, i.e., "any initial point in the space *x,y* approaches the same periodic trajectory in time" (Prigogine 1980, 99). Prigogine also points out (1969, 42) that the differential equations for the Brusselator lead to a solution corresponding to the situation when a uniform distribution of the components *X* and *Y* in space changes into the reverse. That is, there occurs a state when the concentrations *x* and *y* in part of the system differ from those in other parts, and the "system begins to look like a kind of 'factory' with characteristic structural and functional order" (1969, 43). This is the usual manner in which an organized complex structure emerges, where this structure is labeled "dissipative structure" and its becoming in this way is labeled "self-organization."

Which way does a dissipative structure change with the change of the control parameter? The answer comes from the theory of bifurcation, which explores branchings and changes of the types of solutions of non-linear differential equations depending on how the control parameter changes. Prigogine understands the description of the motion of an object via bifurcation as the description of its history. The point is that if we want to interpret a certain state, we need to know the trajectory of the object and the bifurcations that have brought about this state, i.e., we have to know its previous history (Prigogine and Stengers 1984, 161; Prigogine 1980, 106).

The non-linear non-equilibrium thermodynamic theory briefly sketched above is, for Prigogine, the basis for the argument that contemporary science, by employing the above theory, has overcome the standpoint of classical science with its emphasis on the eternal character of laws of nature (Prigogine and Stengers 1984, 10) and has abandoned the idea that reality is simple, immutable and time-independent. According to Prigogine, results of non-equilibrium non-linear thermodynamics appear to prove that reality is historical, time-oriented and complex, thus having overcome the framework of classical science.

7.4.4 *The foundations of the Brusselator*

In order to show the groundlessness of the alleged assertions, let us examine, by means of the categories of rise-origination-history, and evolution-development, the foundations of the Brusselator as a certain type of thought-object.

7.4.4.1 Rise-origination-history and the Brusselator

When describing the processes within the Brusselator it is assumed that, in their totality, they cause no change in the determination of its components. The components X and Y *always* obtain in the system. The scientific law whose equations describe changes in their concentrations should hold always and everywhere, regardless of time. Every bifurcation point (i.e., every point of the bifurcation diagram) represents different solutions for one and the same rate equation of just the same scientific law. It is but a description of purely quantitative changes of the same component of the system. So far as the description through bifurcation is referred to by Prigogine as a theoretical rendering of real history, the latter is recovered, within the theory of bifurcation, by proceeding from an understanding that maintains history always to be a history of one and the same object (e.g. the component X) and to fall under the same law of nature. The theory of bifurcation, as well as the whole account of a dissipative structure's becoming, betrays nothing about the process of the system's origination, described by means of a non-linear differential equation; that process may give rise to a dissipative structure. The system is assumed as already given. By way of illustration of this let me mention the description of the "fight" between cancer cells X and cythotoxic cells E_o, as given in (Nicolis and Prigogine 1989, 224). The processes are as follows:

$$X \xrightarrow{k_1} 2X \qquad \text{(proliferation)}$$

$$X + E_o \xrightarrow{k_2} E \xrightarrow{k_3} E_o + P \qquad \text{(binding and lyse)}$$

Since the proliferation of the cancer cells is non-linear, the whole previously developed apparatus can be applied here. Through the solution of differential equations for the time change of concentrations of the cells, it is possible to describe the "fight" between them as well as to reflect on its possible outcomes. What has to be presupposed here is that the cancer cells are already given; their origin remains here unexplained. The central point of the history of the cancer illness – the initial emergence of the cancer cells – is left unexplained. So, from the point of view of historicity, non-linear non-equilibrium thermodynamics does not transcend the framework of classical science; rather it remains firmly embedded in it.

7.4.4.2 Evolution-development and the Brusselator

How is development-evolution within the Brusselator explained? The processes within the system are described with the help of the equations of a scientific law. Nonetheless it is presupposed that the law that controls the bifurcation parameter is not identical with the law that governs the inside of the system just described. Whether a dissipative structure is to emerge within the system and what type of dissipative structure it may turn out to be are determined not only by the non-linearity of its processes, but also by the ratio of the concentrations of the components *A* and *B*. The situation thus repeats patterns found in mechanics and in other "classical" dynamic theories. To account for a certain state of the described object (e.g., the existence of oscillations within it) one will require both its description via nonlinear differential equations and a description of the surroundings that determine the concentrations of components *A* and *B*. In terms of the temporal succession of processes within the Brusselator, these concentrations bear the character of initial conditions. The same holds for the boundary conditions. It is explicitly assumed from the very beginning that "the system is subject to time-independent boundary condition" (Nicolis and Prigogine 1977, 26). In the case of the Brusselator this implies (since x and y are time-dependent) that the boundary is not a function of the concentrations x and y, which is only possible if the boundary is created and controlled by an external cause, a cause which is completely independent of components *X* and *Y*. The Brusselator as a thought-construct is created as a system with external regulation. Certain regulatory mechanisms are presupposed to exist within the surroundings of the system. Through their own laws they control the concentrations ***a***, ***b*** of components ***A***, ***B*** as well as the space boundaries of the system. That is why the situation is the same as in other dynamic theories. The only end attainable by non-equilibrium non-linear thermodynamics, and by non-linear dynamical theories in general, in their present state will be *at best* an explanation of the system's evolution, through which it responds to environmental challenges. But then the explanation of the becoming of complexity is hardly satisfying. Its becoming is presupposed to be regulated externally (i.e., with respect to the inside of the system) by a blind selection of initial and boundary conditions. The system must from the outside be driven into the bifurcation point ($b = 1 + a^2$) where the dissipative structure comes into being; the system does not organize itself. Therefore, understood in this way, the becoming of an organized structure cannot be labeled "self-organization."

7.4.4.3 Operation-functioning and the Brusselator

Which point of departure underlies the description of the processes at work within the Brusselator, the description of its internal functioning? It is presupposed that, with regard to the system, there exists an external cause responsible for the triggering of the fluxes of components A and B. It is assumed that there exists an external mechanism that pumps matter and energy into the system. The presumption underlying the construction of the Brusselator as a thought-construct is that it is not *causa sui*. The presuppositions-causes of the internal processes – components A and B – need to be independent from the processes inside the system (processes $B + X \rightarrow Y + D$ and $2X + Y \rightarrow 3X$) as well as from the *consequences-effects* of this motion, from components D and E.

7.5 OPEN SYSTEMS WITH INTERNAL REGULATION

Our analysis of Prigogine's approach to the Brusselator reveals that he constructed it as an open system with external regulation. Whence does the idea of external regulation in non-equilibrium nonlinear thermodynamics derive? When recovering the history of thermodynamics (Prigogine and Stengers 1984), Prigogine pointed out that it had covered the path from the exploration of closed systems to the study of open systems. The exploration of the latter led him from inquiring into linear processes that were close to equilibrium to inquiring into nonlinear processes that were far from equilibrium. For all the distinctions in the construal of the object under investigation (closed vs. open; linear vs. nonlinear), they all share a common assumption, namely that the object of inquiry is always a system with external regulation. Prigogine borrowed the idea of external regulation from all the preceding types of thermodynamics, and made it a groundwork for his new thermodynamics. This in turn refutes Prigogine's claim that non-equilibrium nonlinear thermodynamics has already left the framework of classical science.

Marx's approach, as compared with Prigogine's approach, to the object of scientific explanation, shows some profound shifts with respect to classical science. In Prigogine's Brusselator there are, with respect to the essential processes ($B + X \rightarrow Y + D$ and $2X + Y \rightarrow 3X$), two processes of essential formality ($A \rightarrow X$ and $X \rightarrow E$) which are *not* cyclically connected. Contrary to this, in Marx's approach the process of essential formality, as shown above, has a cyclical character. The process of essential formality preceding the essence-producing process and the process of essential formality succeeding the essence-producing process are interconnected. Altogether they set up an operationally closed system which still is open with respect to its environment. Marx's approach does not require the existence of a mechanism, independent from the essential processes in the system, which regulates the input into the system and its space boundaries. The system itself, according to Marx, is the subject of its own inputs – it triggers the fluxes from the environment and it generates and regulates its own space-boundaries. Marx's approach to the object of inquiry, by means of which he transcends the framework of classical science, is that of an *open system with internal regulation*. In the next chapter I will relate the idea of internal regulation to the idea of an internally differentiated essence, and in chapter nine I will relate it to the category of substance-subject and that of the concrete universal.

CHAPTER 8

THE INHERENT LAW AND ESSENTIALISM

The aim of this chapter is to provide the ontological foundations for the scientific inherent, scientific inherent organic, and scientific developmental type of a law. As a starting point I use Kripke's essentialism in its connection to natural kind terms and to the so-called theoretical identification which I briefly lay out. Then, by comparing Kripke's essentialism with B. Enç's essentialism, I show the advantages of an ontologico-epistemologically based essentialism as compared to an essentialism based on Kripke's (1963) formal semantics for modal logic. To stress the difference between these two approaches I compare my ontologico-epistemological approach to scientific laws with the semantic approach pursued by R. Carnap in (1964).

8.1 NATURAL KIND, ESSENCE, AND THEORETICAL IDENTIFICATION

Kripke transfers his understanding of proper names[1] to the understanding of general (common) names (e.g. "gold," "water") which denote entities of a certain natural kind. In his view, these names, like proper names, are rigid designators and therefore lack also a Fregean sense. Any name of a natural kind denotes *one and the same kind* in all possible worlds (where this kind exists). "One and the same" should be understood as entities with one and the same essence independently from their phenomenal properties. In the case of the name "gold" the essence of the objects tagged by this name is the property of having atomic number 79. Even if the knowledge that gold's atomic number is 79 is *a posteriori*, if it is true, then the name rigidly tagged and still tags an element with atomic number 79, independently of the fact if the user of this name knew (or knows) something (or nothing) about the microstructure of the objects so tagged.

As the essential property of an object Kripke takes either its origin or the substance from which it is made (Kripke 1971, 86–87; 1980, 114) and essence is defined as follows (1980, 48):

> When we think of a property as essential to an object we usually mean that it is true of that object in any case in which it would have existed.

All these reflections on the kinds of things, origin, and substance enable Kripke to scrutinize statements of identity containing natural kind terms. They represent theoretical identifications like "Water is H_2O," "Heat is molecular motion," "Lightning is electrical discharge," "Light is a stream of photons." The formulation of these statements is the result of an *a posteriori* form of cognition, but if they are true, then they represent "not contingent . . . but necessary truths, and here of course I don't mean just physically necessary, but necessary in the highest degree" (Kripke 1980, 99). And "statements representing scientific discoveries about what the stuff *is* are not contingent truths but truths in the strictest possible sense" (Kripke 1980, 125). So, according to Kripke, in the case of theoretical identification we are dealing with a *de re* logical necessity cognized in an *a posteriori* manner.

8.2 ONTOLOGICAL OBJECTIONS

Let us start our analysis by scrutinizing Kripke's understanding of "essential property" as given above. Such an understanding is used in the contemporary discussions on essentialism. How one can arrive at it? We do so by reflecting on the essence of a natural kind. Its essence is what makes the kind what it is, by which it differs from other natural kinds. This implies that in all those possible worlds where the respective kind exists, there it has this property, i.e., the essential property is necessary according to Kripke's formal semantics for modal logic. This reasoning, even if perfectly correct, bears in itself a fundamental shortcoming. It states only that if a property is essential for a natural kind, then it is necessary for it; but it does not say when a property is essential for a natural kind. Another shortcoming of Kripke's understanding of essential property is that the reverse of "if a property is essential for a natural kind, then it is necessary for it" does not hold.[2] A property which is necessary according to Kripke's formal semantics for a natural kind, i.e., which is instantiated in all possible worlds where entities of this kind exist, need not be an essential property. The class of necessary properties of a natural kind can be broader than the class of its essential properties.

Let me take as an example of this the case, introduced in subchapter 3.1, of the difference between the organic composition of capital in the branch turning out the commodity x, ORG_Bx, and the organic composition of capital in the whole economy, ORG_Ex, of which the single branch is an element. Marx denotes by "organic composition of capital" the ratio between the variable capital (buying labor force) and the constant capital (buying the thing-components of the production process). Their ratio is determined by the implementation of the results of technical progress, by means of which living labor is replaced by "dead" labor. This progress, as well as its implementation, takes its course in a certain branch independently from its course in other branches of production. The state where in each branch of production a unit of variable capital would correspond to the same number of units of constant capital is not possible. In each possible world where there exists capitalist economy, its branches have different organic compositions of capital. This property of the capitalist economy is, from the point of view of the formal semantics of modal logic, a necessary property but one which is still not an essential property. By "$ORG_Bx - ORG_Ex \neq 0$" we denote a non-essential – contingent quantitative difference between qualitatively identical entities – i.e., entities identical with respect to their essence – which mutually interact.

Kripke's notion of *de re* logical necessity also suffers[3] from a lack of clarity. Let me first take his example "Hesperus is Phosphorus." Kripke claims that the fact that Venus is observed in the evening in a certain position and the fact that it is observed in the morning in a certain position are only "the contingent marks by which we identify a certain planet and give it a name" (1980, 105). To find out what Kripke understands in this case of identification as *de re* logical necessity, let me first abstract from these contingencies brought about by the existence of cognizing beings. Let me suppose that human beings never came into being on Earth, and so nobody observed, observes, or will observe Venus in the morning (and tag it "Phosphorus") and nobody observed, observes, or will observe Venus in the evening (and tag it "Hesperus"). In such a case, instead of "Phosphorus is Hesperus" we have to write "Venus is Venus." But then, what is the *de re* logical necessity given in the very objective reality? It is definitely not that which can be symbolized as $N_l(x)(x = x)$, where "N_l" stands for logical necessity.

Let me take another example of Kripke's, namely "Heat is molecular motion." In this case, according to Kripke, the phenomenon tagged "heat" is independent from the existence of human beings. Even if humans had never come into being on Earth, this phenomenon would still exist, even if it would not produce in us the sensation of heat. We get acquainted with heat in an *a posteriori* manner through sensation, and we do not know *a priori* the phenomenon which produces this sensation in us. According to Kripke, only after a certain time do we discover "that this phenomenon is in fact molecular motion. When we have discovered this, we have discovered an identification which gives us an essential property of this phenomenon. We have discovered a phenomenon which in all possible worlds will be molecular motion" (Kripke 1980, 133). He claims that the reduction of the phenomenal properties to the essential property has the character of an *a posteriori* discovery of something which is identical with these phenomenal properties and claims also that "temperature is identified with mean molecular kinetic energy" (1980, 129).

The equation which implicitly appears here is $u = 3kT/2$. Here u denotes the mean kinetic energy of a molecule of an ideal (one-atomic) gas; k denotes Boltzmann's constant; T denotes absolute temperature. If temperature would be completely reduced to molecular motion, i.e., both would be of the same quality, then the constant k mediating in the equation between heat and molecular motion would be, as proved already in chapter six, dimensionless. But Boltzmann's constant unifies in itself the (derived) unit of Joule and the unit of Kelvin. It is not dimensionless and mediates here between two different qualities. The problem which kinetic theory faces is that even if it views temperature of a gas as a macroscopic phenomenon which has its ground-cause in the kinetic energy of the chaotic movement of particles of this gas, still it has to approach that ground via the magnitude of temperature. Temperature as it is given in the equation $u = 3kT/2$, is the *external measure* of the kinetic energy of the chaotic movement of the gas particles. Along with this, it is noteworthy that in the kinetic theory the existence of the gas does *not* follow from the fact of the existence of molecules. While deriving in this theory the above-mentioned equation it is explicitly presupposed that we are dealing with microobjects which, with the exemption of short-lived collisions, do not interact at all. The very existence of the gas is a result of action which with respect to the essential property – molecular motion – has a completely non-essential, accidental character. So, even if it holds

In all possible worlds where macroobjects with properties like temperature, pressure, volume, etc., exist, there also exist (in a certain way) microobjects with kinetic energy,

it does not hold

In all those possible worlds where microobjects with kinetic energy exist, there also exists (in a certain way) a macroobject with properties like temperature, pressure, volume, etc.

Because of this, the equation $u = 3kT/2$ does not state that absolute temperature and mean kinetic energy are identical. It states that the temperature is a macroscopic phenomenon which has its ground in the unorganized microscopic motion of the smallest particles. The kinetic theory of gases states that this (microscopic) motion appears at the macrolevel as temperature. So, while Kripke claims that in theoretical identifications we discover something which is identical with a certain class of phenomena, this example shows that in the case of theoretical identifications we discover some "being for itself," something independent from and behind the phenomena. So, e.g., in the case of the equation $u = 3kT/2$ we discover that everywhere the phenomenon of temperature exists, there will exist behind it – as its *ground-cause* – the movement of molecules.

8.3 THE ESSENCE AS THE GROUND OF A NATURAL KIND'S PHENOMENA

If our analysis of Kripke's statements on theoretical identifications is correct, then we face a problem which can be stated in the form of the following question. Where does the *de re* logical necessity reside if not in the identity of the phenomenal properties with an essential property? A partial solution is given in B. Enç's papers (1975) and (1986), which are characterized by a strict conceptual distinction between necessary (in the sense of Kripke's formal semantics for modal logic) and essential properties, as well as by a non-identificational approach to the phenomenon-essence relation. He introduces the following distinctions (1975, 97–98):

[i] Of the whole set of properties of a given kind of thing a subset will be governed by laws of nature that are about things of that kind. I shall call these the *natural properties* of that kind of thing.

[ii] Another subset will be shared by all the individuals of that kind and yet the fact that that kind of individual has those properties will not be a law of nature that governs things of that kind. I shall call the second subset the *factitious properties* of that kind.

[iii] If we examine the natural properties of kinds we notice that a very small subset of them . . . are causally responsible for the whole set. This small set will be called the *essential properties* of that kind of thing.

[iv] All those properties that are explained in terms of essential properties will be called *necessary properties* of that kind of thing. They are necessary in the sense that they are natural consequences of the essence of that kind of thing.

Accordingly, Enç explicates the term "necessary properties" in a twofold way: in relation to the term "essence," and in accordance with logical semantics. In the latter case he replaces it with the term "factitious properties." He then defines necessary properties as follows (1975, 99):

Necessary properties of a kind of thing are those and only those . . . which follow, via laws of nature, from essential properties of that kind of thing.

Enç further develops his views in (1986) by reflecting on the relations of events, as well as on relations in which "an object's having a certain structural property constitutes its having some other (manifest) properties" (1986, 405). He labels both relations with the expression "makes it the case that" or simply as *M*. As its illustration he suggests the following example. Hydrochloric acid has a great number of properties: it is clear; it has a distinctive odor at room temperature; it solidifies at *t* degrees Celsius, etc. Let me denote these properties as $P_1, P_2, \ldots, P_n$. In addition it has also a great number of causal consequences: if ingested, it kills; if lead is placed in it, the lead dissolves, etc. Hydrochloric acid also "has a distinctive molecular structure," where this "is essential to it, that is, in any possible world in which a pool of liquid is hydrochloric acid, that pool has that particular molecular structure; if the molecular structure is altered, it will *cease* to be hydrochloric acid" (Enç 1986, 409).

Let me now consider all those possible worlds in which hydrochloric acid exists. In some of them there exist conditions which are different from the conditions of the actual world which enable the becoming of the properties denoted as $P_1, P_2, \ldots, P_n$. Thus, for example, "in a possible world in which there is *schmight* instead of light, and where *schmight* is the medium that makes vision possible, except that it is reflected by objects that transmit light, that body of hydrochloric acid will not be clear" (Enç 1986, 409). Or, in a possible world in which the atmospheric pressure is much higher than in the actual world, the acid would not vaporize and would so lack its typical odor at room temperature. The same holds for the causal instances of the *M*-relation. The causal characteristic of "deadliness" – that hydrochloric acid kills if ingested – obtains only in that set of possible worlds where not only hydrochloric acid exists but also an organism with a certain type of metabolism and a certain type of digestive system exists. Enç then explicates the *M*-relation as follows (1986, 409–410):

a's being *F* is what makes it the case that *b* (or *a*) is *G* if
(i) *a* is of kind *K*
(ii) *F* is a component of the property that essentially defines *K*; and
(iii) each possible world in which *a* is (has) *F*, and in which the relevant kinds of the actual world obtain, that world contains the event (or state of affairs) in which it is true that *b* (or *a*) is *G;* and
(iv) each possible world in which *a* is not (fails to have) *F*, and in which the relevant kinds of the actual world exist and the relevant conditions of the actual world obtain, that world does not contain the event (or state of affairs) in which it is true that *b* (or *a*) is *G*; and
(v) in any possible world in which any object of whatever kind is *F*, and in which the relevant conditions hold, there exists some object that is *G*, but it is not the case that in every possible world in which any object of whatever kind is *G*, there exists some object that is *F*.

If I sum up Enç's reflections, I come to the conclusion that he analyzes the relation of the essence-ground-cause to its own manifestations-effects, which come into being under certain (non-essential conditions). So, he understands the essence of a natural kind, contrary to Kripke, as the *ground of the manifestations of this natural kind*; as the *internal of the external.*

8.4 THE ESSENCE AS THE INTERNALLY DIFFERENTIATED GROUND OF A NATURAL KIND'S PHENOMENA

On the one hand, Enç's reconstruction of the *M*-relation already shows that the term "essence" can be substantiated without and prior to the introduction of the term "necessity" in the framework of logical semantics. But on the other hand, this substantiation still is unable to locate the domain of Kripke's *de re* logical necessity. The *M*-relation stands for the relations of the essence of a natural kind to its manifestations which are mediated by non-essential conditions. Thus, the relation denoted as M cannot have the character of a *de re* logical necessity.

This enables us to revise Ruben's view, analyzed in subchapter 2.7, on the logically necessary relation between two properties of a natural kind. While he claimed that such relation does not hold "if these properties are had only non-essentially by things of kind N," (1979, 64) I now can formulate a weaker statement. If of two properties of a natural kind at least one is a non-essential property of this kind, then their relation has not the character of a logically necessary relation even as applied to that natural kind.

To locate Kripke's *de re* logical necessity, let me first compare Enç's ideas about the M-relation with his own classification of laws of nature (1986, 410):

> It is assumed that the fundamental laws of the actual world governing things of kind K hold in all possible worlds that are being considered here. But since not all the kinds of the actual world are assumed to exist in each possible world containing Ks, some of the so-called derived laws of the actual world may not hold for them.

From points (i) through (v) characterizing the M-relation stated earlier, it follows that the derived laws of nature regulate the relation of kind K's essential property F to the nonessential property G. But Enç does not explain what relations are regulated by the fundamental laws. A possible solution can be found by means of Enç's hydrochloric acid example. Here (1986, 409) he claims that it

> seems plausible to expect that for any P_i of the set of Ps, there is some component of the molecular structure that is essential to hydrochloric acid, which makes it the case that a given pool of hydrochloric acid has that P_i.

From the view of the characterization of the M-relation by points (i) through (v), we have here the essential properties $F_1, F_2, \ldots, F_n$ of kind K, which under certain conditions manifest themselves as $G_1, G_2, \ldots, G_n$. Even if Enç presupposes that the very essence is determined by a certain number of different properties – that it is internally differentiated – he draws conclusions only about the relation between F_i and G_i – only about the M-relation of the essence (internal) to its manifestation (external) – but never about the relation between F_i and F_j, i.e., never about the relation of the internal to internal.

In my view, fundamental laws regulate the relations between the very essential properties, and it is the mutual relation of the essential properties of a natural kind which has the character of a *de re* logical necessity. This relation can be explicated as follows:

(i*) If F_i and F_j denote the components of a property which essentially defines kind K (i.e. they set up the ground of this kind), then
(ii*) in all possible worlds where certain a is (has) F_i it is (has) also F_j; and
(iii*) in all possible worlds where a is (has) F_j it is (has) also F_i.

Let me label the relation of the internal to the internal as an I-relation. Its explication in (i*) through (iii*) enables me to redefine the term "essence of a natural kind." Earlier I claimed, following Enç, that the essence of a natural kind is the ground of the manifestation of this kind. Now I claim that the essence of a natural kind is the *internally differentiated ground of the manifestations of this kind.*

Such an understanding of the essence of a natural kind overcomes, in my view, its understanding as given in all the contemporary discussions on essentialism. The latter still remain at the stage of analysis of the relation of internal to external, never arriving at the analysis of the relation of internal to internal.

The fact that I label the relation of the phenomena (appearances and manifestations) to the essence also by the terms "internal" and "external" enables me to characterize the typology of scientific laws expounded in this book as follows:

(i) the empirical type of a scientific law, L_e, expresses the knowledge of the relation of the external (appearance) to the external (appearance);
(ii) the pure idealizational type of a scientific law, $L^{(k)}$, expresses the knowledge of the relation of the internal to the external (the appearance);
(iii) the inherent types of a scientific law: L_i, L_{io}, L_d, L_h, express the knowledge of the relation of the internal to the internal;
(iv) the inherent idealizational type of law, $L_{ii}^{(k)}$, expresses the knowledge of the relation of the external (manifestation) to the internal.

8.5 THE ONTOLOGICAL FOUNDATIONS FOR L_i, L_{io} AND L_d

Let me now analyze the relation of the essential conditions to the essence and to the essence-producing substance, as it is expressed in scientific laws of the type L_i, L_{io}, and L_d. On the one hand, each of the l essential conditions by itself is causally necessary for the initiation of the essence-producing process, so that the essence comes into being. On the other hand, none of these essential conditions causally suffices by itself for the initiation of the essence-producing process. Only their unification makes up the *cause* which is causally necessary and causally sufficient for the existence of the substance and for the production of the essence from it. The essence is here the *effect*; it is produced as the result of their unification. The relation of the essential conditions to the essence is here such that in all possible worlds where certain essential conditions exist, there will also exist a certain essence-producing substance and the corresponding essence will come into being. Scientific laws of the type L_{io} at the same time show that in all possible worlds where the essence is produced from the essence-producing substance, the essential conditions will be renewed, reproduced.

The conclusion we arrive at, therefore, is that in the case of scientific laws of the type L_i, L_{io}, L_d we deal with an *a posteriori* cognized *de re* logical necessity. While in the previous subchapter I claimed that it resides in the internally differentiated essence, now this claim can be rendered more precise by saying that it resides in the mutual relation of the essential conditions, essence-producing substance, and the essence.

A possible objection[4] to this view might be that the necessity given here is not a logical one because it is a matter of pure accidence which concrete essential conditions exist, and so what kind of essence exists at all. This objection does not hold because it does not distinguish between two aspects of the *de re* necessity. The first aspect of *de re* logical necessity, as it is fixed in scientific laws of the type L_i, L_{io}, and L_d, pertains to the relation between the essential conditions, the essence-producing substance, and the essence, if they are already given at least one time. In all possible worlds where they are given at least one time, everywhere there and only there they mutually produce and reproduce each other; their relation is a logically necessary one. The second aspect of the *de re* logical necessity pertains to the *necessity of the historical becoming* of the essential conditions. While scientific laws of the type L_i, L_{io}, and L_d express the necessity of the *production*, *reproduction*, and *development* of an object which has already become, scientific laws of the type L_h state if[5] the essential conditions had to come into being even one time through previous development.

Our reconstruction of the ontology connected to scientific laws of the type L_i, L_{io}, and L_d enables us to find an answer to two as yet unanswered questions which were formulated in connection with the ontology of the scientific law of type $L^{(k)}$ in subchapter 4.3. The first question was as follows: What is the relation between the membership of an individual to a kind *N* and the fact that it has the essential property *H*? By amendment of the logical structure characterizing scientific laws of the type $L^{(k)}$ we arrived at a counterintuitive conclusion that there can exist an individual with the essential property *H* but which is not from kind *N*, even if *H* is the essence of this kind. This conclusion is the result of the fact that in scientific laws of the type $L^{(k)}$, the ground *H* of kind *N* is not grasped. The essence is grasped through a relation with the structure $f_1(H) = E^{(k)}$ only as an identity of phenomena-appearances, as something to which we in the long run converted all the phenomena, without comprehending what this very ground is. That relation signals that the kind *N* was not completely in mind converted and reasoned by means of its own ground *H*. Only in scientific laws of the type L_i, L_{io}, and L_d is the natural kind completely converted to its own ground by showing how the kind is created and functions primarily through the production, functioning, and development of its own essence-ground. For each individual, characterized by the essential conditions Cim_{1-l}, the property *H* is necessarily produced from the essence-producing substance, and through this production the individual creates the ground of the natural kind ("pure" kind) as well as its own membership in it.

Working from this logic, we find that there exists another important difference between laws of the type $L^{(k)}$ and those of the type L_i, L_{io}, and L_d. As we have already shown in subchapter 4.2, in cognition based on scientific laws of the type $L^{(k)}$ one proceeds from the concept of a natural kind,[6] which is *here* fixed – given through the cognized superficial properties[7] – to the cognition, concept of essence. The concept of natural kind is here *primary* and the concept of essence *secondary*. This situation is completely reversed by the passage of cognition to a level where inherent types of scientific laws are formulated. The concept of the natural kind is converted, one might say "dissolved" in its principal determinations: essential conditions, essence-producing substance, and essence. In knowledge based on inherent types of a scientific law, it is possible by means of the concept of essential conditions, essence-producing substance, and essence to show how their relation creates the ground of the very natural kind. At this level of knowledge the concept of natural kind appears, with respect to the concept of essential conditions, essence-producing substance, and essence, as a *derived, secondary* concept. The concept of natural kind appears as a secondary concept in a cognition based on inherent types of laws also because, by the movement of cognition from a scientific law of the type L_i towards laws of the type $L_{ii}^{(k)}$, $L_{ii}^{(k-1)}$, . . ., $L_{ii}^{(k-j)}$, one explains how the *kind is created* as an internally differentiated, acting whole, as a system of *mutual interaction*. Such a system-whole I will, following Marx (1981a, 35; Marx and Engels 1975ff., Vol. 28, 37) label "organic system." So we can state that by the movement of scientific cognition from laws of type $L^{(k)}$ to laws of type L_i, L_{io}, L_d, and $L_{ii}^{(k)}$ the category of natural kind is deepened into the category of organic system.

This chapter, together with chapter four through seven, reveals that Marx's *Capital*, viewed as an explanatory structure, represents the exemplification of three categorial clusters. The first volume corresponds to the cluster which I label "The essence-producing process in an organic system"; the second volume corresponds to the cluster "Process of essential formality and its place in the organic system"; finally the third volume corresponds to the cluster "The organic system as a whole." In chapter eleven I will compare these three clusters with the clusters "Essence," "Manifestation," and "Reality" from Hegel's second book *Essence* of his *Science of Logic*.

The second question which appeared in the reconstruction of the ontology of scientific laws of the type $L^{(k)}$ in subchapter 4.3 was as follows: Why entities from one natural kind have one and the same essence? Is it not just a matter of linguistic convention? This question could not be answered in the categorial "outfit" of cognition based on scientific laws of the type $L^{(k)}$ because in laws of this type, the concept of natural kind is fixed by means of nonessential-apparent properties and not by means of essential properties. The very concept of an essential property appears here only as a sense of a term by means of which *we* tag that what *we* discovered as common to the manifold phenomena, and it was not clear at all if something in reality corresponds to it, something which has a self-being, which can be investigated and explained independently from phenomena (appearances and manifestations). This fact forces us to formulate this question even if it is evident that what unifies individuals into one natural kind is only this: that each of them has the essential property which sets up the ground of this natural kind (the "pure" natural kind).

So, if we think about the relation of a natural kind to its essence, it is necessary in the long run to proceed from the concept of natural kind to that of essence. The above-mentioned question is answered in the framework of cognition based on inherent types of a scientific law as follows. If certain individuals are characterized by the same essential conditions, then one and the same essential property is produced in them, while simultaneously the ground of one and only one natural kind comes into being, and so also the membership of these individuals in one and the same natural kind comes into being.

The explication of the type of necessity fixed in inherent types of scientific laws enables us also to define completely the concept of modification conditions. While distinguishing them from singular conditions, I claimed that the latter can be predicated for individuals which exist in more than one possible world and which are denoted by individual variables. This understanding can now be made more precise by saying that the modification conditions can be predicated for individuals existing even in all possible worlds, i.e., they can be *necessary* as defined by Kripke's formal semantics. The essential conditions can also be predicated for more than one possible world, even for all possible worlds. But this necessity is an *essential* one, i.e. tied through a *de re* relation to the essence-producing substance and the essence. The necessity of the modification conditions is a nonessential one; it is a contingent necessity. So, I arrive at a differentiation of two senses of the term "necessity." One sense is given by Kripke's formal semantics of modal logic, while the second is implied by the understanding of the essence as an internally differentiated ground of manifestations. The necessity of the modification conditions I characterize as a contingent one because the term "contingency" I take as a pair term not in relation to the term "necessity" explicated in Kripke's formal semantics, but as a pair term in relation to the term "essence."

8.6 THE ONTOLOGICO-EPISTEMOLOGICAL vs. THE SEMANTICAL APPROACH

The character of my approach to scientific laws can be made apparent by extensively citing from Carnap's approach to causal modalities in (1964)[8]

Carnap, using his semantic approach to the explication of logical modalities in (1956), claims (1964, 951–952)

The concept which I would try to explicate first is the semantical concept of the causal validity of a sentence, corresponding to the modal concept of the causal necessity of a proposition, in other words, the distinction between the genuine causal universals (laws) and accidental universals. . . . For this purpose it seems to be essential to distinguish between fundamental and derivative laws. *Derivative laws* may have the same semantical nature as accidental universals; therefore we cannot distinguish them directly by a purely semantical analysis of their forms. On the other hand, I believe, although I am not quite certain, that it is possible to explicate the distinction between *fundamental laws* and accidental universals in purely semantical terms.

The central problem so far unsolved is that of explicating the concept of sentences of *nomic form* . . . If this problem were solved, I would proceed as follows. I define:[9]

(2) A sentence S in L is a fundamental law in $L =_{Df} S$ has nomic form and is true in L.

(3) A sentence S in L is *causally valid* in $L =_{Df} S$ is L implied . . . by the class of the fundamental laws in L.

Note that the concepts defined in (2) and (3) are semantical concepts which must be clearly distinguished from corresponding . . . inductive and epistemological concepts. . . For the explication of causal modalities we must take the former, not the latter. . . . Let S be a sentence in a language of physics with a space-time coordinate system. It seems also clear that the following is a necessary but not a sufficient condition for S having nomic form: S must not contain space-time constants, but only variables, as J. C. Maxwell has first pointed out.

My approach differs from that of Carnap as follows. Firstly, concerning the issue of a sentence of nomic form, i.e., the criteria a sentence has to fulfill to be a potential candidate for a scientific law, this is not an issue that should be decided by categorically accepting certain sentences and rejecting others; this is not an all or nothing issue. By differentiating between the empirical type, the pure idealizational type, and inherent types of scientific law, I have come to interpret the issue of nomicity as an issue of degrees. Each of these types is a type of a scientific law in its own right, differing from the others by its respective epistemic depth. Their difference is primarily a difference in their respective "outfit" provided by the epistemico-ontological categories.

Secondly, my differentiation of the degrees of epistemic depth is based on a non-Humean approach to causation. This enabled me to claim in chapter eight that any of the inherent types of a scientific fundamental law theoretically expresses the relation between essential properties of a natural kind; theoretically grasps the *I*-relation. Given my characterization of the inherent types of a scientific law, I now claim that each of them represents a fundamental type of scientific law. Having also in mind my concept of the essential (inherent) conditions, appearing in the formulation of the inherent types of scientific law, it is readily seen that my understanding of a fundamental law differs essentially from that of Carnap. While he claims that a fundamental scientific law should be absolutely independent from any conditions prevailing in the universe, I claim that a fundamental scientific law, while independent from modification conditions and singular conditions, still depends on inherent (essential) conditions. So, while Carnap understands Maxwell's criterion of law-likeness[10] as independence from any conditions, I reinterpret it as independence only from modification conditions and singular conditions. This reinterpretation is enabled by the introduction of yet another criterion of law-likeness, namely the criterion of dependence on inherent conditions. Given the fact that the idea of essential (inherent) conditions and of inherent types of a scientific law goes back to Marx, I label it, following Černík's (1977), as *Marx's criterion of law-likeness*.

Thirdly, my non-Humean approach to causation enables me to differentiate between fundamental laws and derivative laws in a way which completely differs from Carnap's approach. While he sets the criterion of differentiation between derivative scientific laws and fundamental scientific laws as dependence vs. independence on any conditions, I put forward the following criterion. A derivative law, as shown already in previous chapters, expresses the relation between the essential and manifest (non-essential) properties of a natural kind; it theoretically grasps how the *I*-relation manifests itself, under non-essential conditions, as the *M*-relation. I claim, therefore, that while fundamental laws of science depend only on inherent conditions, a derived law depends both on essential and non-essential conditions. Given my characterization of the inherent idealizational type of a scientific law in chapter five, one can readily see that it represents what is usually labeled a derived type of a scientific law.

Fourthly, the non-Humean approach, contrary to the Humean regularity approach, enables me to put forward a criterion for the differentiation between derivative scientific laws and accidental universals. The regularity approach correctly grasps that derived scientific laws depend on certain ("local") conditions. But simultaneously – on the basis of Maxwell's criterion understood as independence from any conditions – it wrongly claims that an accidental universal holds only locally (due to local conditions). Therefore, the regularity approach cannot put forward a criterion for that differentiation. In my view, based on the reinterpretation of Maxwell's criterion of law-likeness by means of Marx's criterion of law-likeness, the real (ontical) differentiation is the existence vs. non-existence of an underlying generative mechanism. On the one hand, a derivative law (an inherent idealizational type of a scientific law) represents already a theoretical knowledge of such an underlying generative mechanism. On the other hand, a Humean causal law (a phenomenological type of a scientific law), as shown in chapter one, does not as yet constitute a law that grasps this underlying mechanism. Therefore, regarding knowledge there is no epistemic difference between an accidental universal and a phenomenological type of a scientific law. When science initially, at the level of empirical generalization (in the words of C. G. Hempel), puts forward a universal fixing a regularity relation between phenomena, one never knows if behind this regularity there exists, as its ground-cause-essence, a certain underlying generative mechanism. If there is no such mechanism, then the universal has a purely accidental character even if this universal, sometimes, holds on a cosmic scale.

Fifthly, all my differentiations between fundamental laws, derivative laws and accidental universals are based primarily on epistemico-ontological considerations. The latter set up a basis upon which one can *then* realize, by means of symbolic logic and logical semantics, an investigation of the language of scientific laws. This is, for example, readily seen in my reconstruction of the inherent type of scientific law in Chapter 5. By reconstructing its epistemic (cognitive) structure I open the way for a future syntactical and semantical investigation of the sentential connective appearing in the language expressing it, symbolically represented by me as "$\xrightarrow[T]{}$", whose tentative reading in English can be given as "If . . ., then . . . is necessarily produced, so that . . .". So, while Carnap proposes to explicate causal modalities by starting from semantical considerations, and then proceeds to epistemological considerations, I propose to take the other way round. Of course, philosophy of science is now in a more favorable position than it was forty years ago. For it is a well-known fact that until now all attempts to explicate the concept of scientific law by means of symbolic logic and logical semantics have failed.

CHAPTER 9

SUBSTANCE-SUBJECT

The previous chapters showed that some of the central issues in the philosophy of science, like regularity, development, becoming (origination) and decline, can be solved in a framework of categories by means of which one reconstructs the inherent organic, the developmental, and the historic type of scientific law. In each of them the idea of *reproduction* has a central position. If one analyzes this idea from the point of view of the cause-effect relation, he or she finds that in the process of reproduction the cause – the essential conditions – and the effect – the essence-producing process – continually change place. Each of them is both a given supposition as well as result of the other. The process of reproduction as a causal process has the character of a *causal loop*. According to Marx (1982a, 2243; Marx and Engels 1975ff., Vol. 34, 207–208)

> [t]he production viewed as a process continuously renewing itself, or viewed in the connection (*Zusammenhang*) of its continual (*stetigen*) renewal is *reproduction*. . . . The suppositions of production appear themselves in the *reproduction* as its past results, and the result of production as its *supposition*. In each reproduction each supposition appears as a result (setting (*Setzen*)) and each result as a supposition.

The aim of this chapter, therefore, is to provide an analysis of causal loops, or of the phenomenon usually, and as I will show misleadingly, labeled "backwards" causation. Contrary to previous chapters, the empirical counterpart of and the testing ground for my philosophical claims in this chapter will be not only Marx's works on political economy, but also the work of scientists like Manfred Eigen who deal with the origin of life on Earth and especially on the issues related to the so-called hypercycle (Eigen 1971; 1976; 1980; 1983; 1986; 1993a; 1993b; 1994; Eigen et al. 1981; 1989; 1991; Eigen and Schuster 1979; 1982; Eigen and Gardiner and Schuster 1980; Eigen and McCaskill and Schuster 1988; Eigen and Winkler-Oswatitsch 1981a; 1981b; 1981c; 1992; Eigen and Winkler-Oswatitsch and Dress 1988; Oswatitsch-Winkler and Dress and Eigen 1986; Schuster 1981).

The analysis of causal loops and the reconstruction of the category of substance-subject based on it will then serve as a starting point for the analysis of the issues related to the problems of universals.

9.1 MELLOR, MACKIE AND REICHENBACH ON "BACKWARDS" CAUSATION

Mellor's analysis of causal loops in (1981) is a part of his reflections which try to show the impossibility of "backwards" causation, the latter being usually understood as an effect which is temporally earlier than its own cause. His reflections are based on the following two assumptions.

1. Causes make their effects more likely (probable) than they would otherwise have been (1981, 123).
2. The source of time's direction lies in the causal order of events (1981, 144–145).

Because of the second assumption, Mellor views "backwards" causation not only as a time-reversal, but also – and primarily – as the *reversal of the very causal order*. This in turn leads him to the following claim ". . . the only way I can conceive backward causation and backward time travel occurring: it requires a closed chain of causes and effects" (1981, 176–177). The precondition for the existence of "backwards" causation is the existence of causal loops. "Unless they form such a loop, I see no basis for saying that the causal and hence the temporal orders of two pairs of events oppose each other" (1981, 177). But because of the first assumption given above (that causes make their effects more probable than they would otherwise have been) the idea of causal loops, according to Mellor, leads to conclusions refuting it. Let us take the singular[1] event e and the event e^*, so that if there is a causal loop-like relation between them, then e causes e^* and *vice versa*. Let $P(e^*|e)$ stand for the probability of the singular event e^* to occur given the fact that the singular event e occurred, while $P(e)$ and $P(e^*)$ stand for the unconditional probability of the event e and e^*, respectively. Suppose now that $P(e) = n$ holds $(0 \leq n \leq 1)$ and suppose also that e in fact occurred. Then the relation $P(e^*|e) > P(e^*)$ holds and e^* occurred as well. But, in this case, not only does e cause e^* but also *vice versa*, $P(e|e^*) > P(e)$ holds as well. But this means that the probability of the event e, in the case when it is the *effect* of the singular event e^*, should be greater than its probability when it is the *cause* of the singular event e^*; i.e., one ends up with two mutually contradictory claims: $P(e) = n$ and $P(e) > n$. So, according to Mellor, the idea of a causal loop leads to the conclusion that singular events have mutually exclusive properties. Therefore, (1981, 11–12):

> there will . . . be no closed timelike, i.e., causal, loop from action to effect . . . and back to the action . . . whatever happens the causes and the effects will form no such loop . . . The correlation which causation needs cannot be attributed to it all the way round a closed chain of events.

Mackie's approach to causal loops in (1980) is not only, like Mellor's, based on his reflections on "backwards" causation, but also on teleology. According to Mackie, one possible approach to the understanding of the direction of causation is (1980, 168)

> by way of the notions of control and 'effectiveness': the direction of causation is connected with the way round in which we might use a causal law as a recipe. A cause is *effective* as a means of bringing about its effect . . . But this brings out one of the queerest aspects of the hypothesis of backward causation. One might suppose that if there were a time-reversed exact analogue of causation, with causal priority opposite to temporal priority, then one could use a later event as a means to bring about an earlier one . . . If there were such a thing as backward causation, it would somehow have to stop short of offering us means of bringing about the past.

Mackie's "worries" about backward causation are based on Scriven's (1957) well known example of a precognizer who on Monday produced a drawing of a picture which was put in front of him on Tuesday of the same week.

Mackie then continues with his analyses of backward causation by means of the concept of final and efficient causation, which he views as related to the issue of teleology. A teleological or final cause means "that something – perhaps an event, or a state, or a fact – is an end or goal to which something else is, or is seen as, a means" (1980, 272). Even if it is wrong usually to explain the occurrence of a cause by referring to its effect, in the case of the existence of some sort of backward causation, as in the Scriven example, the later event would explain the earlier event because the cause would become fixed by its own effect. In such a case "we could call the later event an end or goal simply in virtue of its relative temporal position, rather (as we have been doing) a time-reversed cause" (1980, 274). Unfortunately, according to Mackie, "[t]he only trouble is that such cases seem never to occur" (1980, 275).

The same claim is made by H. Reichenbach (1956). According to him (1956, 36):

> [t]ravelling along causal chains we . . . make the discovery that we never return to the starting point; or, to put it another way, that *there are no closed causal chains.*

He represents a closed causal chains by the following scheme:

Figure 4.

and accompanies it with the commentary that "[a]rrangements of this kind never occur" (1956, 39).

In my view Mackie's claim that such cases *never* occur is based on trivial cases, like Scriven's case of precognition, and not on an analysis of what special (natural and social) science given in his time claims about causal relation in reality. The same holds also for Mellor's reasoning on causal loops (1981, 177). The latter is based on Dummett's (1964) paper which dealt with the case of a person claiming at time t_1 the ability to ignite a fire an hour earlier (at t_0) by putting to it a match an hour later (at t_2).

Examples upon which Mackie and Mellor based their reflections on causal loops were perhaps justified thirty years ago, but definitely not since the early seventies. The fact is that three years before the first publication of Mackie's *The Cement of the Universe*, Manfred Eigen published his (1971) paper which addressed the issue of the origin of life by *explicitly reflecting on the cause-effect relation in a loop-like manner.*

Its introduction, under the caption "cause and effect" starts with the following claim (1971, 465–467):

> The question of the origin of life very often appears as a question about 'cause and effect.' Physical theories of macroscopic processes usually involve answers to such questions . . . It is mainly due to the nature of this question that many scientists believe that our present physics does not offer any obvious explanation for the existence of life, which even in its simplest form always appears to be associated with complex macroscopic systems such as the living cell. As a consequence of the exciting discoveries of 'molecular biology,' a common version of the above question is: *What came first, the protein or the nucleic acid?* The term 'first' is usually meant to define a causal rather than a temporal relationship, and the words 'protein' and 'nucleic acid' may be substituted by 'function' and 'information.' The question in this form, when applied to the interplay of nucleic acids and proteins as presently encountered in the living cell, leads ad absurdum because 'function' cannot occur in any organized manner unless 'information' is present and this 'information' only acquires its meaning *via* the 'function' for which it is coding. Such a

system may be compared to a closed causal loop. Although it is evident that the line from which the loop is formed must have originated somewhere, the starting point will have lost all its importance as soon as the circle is closed. The present interplay of nucleic acids and proteins corresponds to a complex hierarchy of 'closed loops' . . .

9.2 THE HYPERCYCLE

There are two problems which any scientific theory aiming at the explanation of the laws regulating the becoming and evolution of life on Earth has to solve. One is the fact that life, even in the most elementary forms in which it exists today, represents such a complex phenomenon that it could not originate in a purely accidental way.

If one takes, e.g., the four bases of the nucleic acids – adenine, guanine, cytosine, and thymine (uracil) for DNA (RNA) as "symbols," one finds that a gene ("sentence") of the length N represents 4^N different sequences. For a gene of $N = 1000$ it means $4^{1000} = 10^{602}$ of equally possible variants.[2] For life on Earth to be a product of pure chance, all these possible variants had to be realized; but this is not possible "within the spatial and temporal confines of our planet" (Eigen – Winkler-Oswatitsch 1992, 10).[3] On the assumption that (1992, 10–11)

we covered the earth with a layer of solution one centimeter thick, containing nucleic acid at a concentration of one gram per liter, and allowed the nucleic acid molecules in it to form and decay with a life of not more than one second each, then after one thousand million years there could have arisen 10^{50} fresh molecules,

and that is a far cry from the above required 10^{602} variants. The origin of life could not be the result of a random process, but was an optimized process regulated by certain laws or principles. The discovery and formulation of these principles in a scientific theory should also enable us to solve the second problem of the origin and evolution of life. Even if Earth is populated today by a huge variety of species, animals and plants (*diversity of evolution*), still "there is only one basic molecular machinery of the cell: one universal genetic code and unique qualities of the macromolecules" (*unity of evolution*) (Eigen and Schuster 1979, 1).

By applying the theory of games (Eigen and Winkler 1974; 1981a) Eigen formulated the principles of evolution as follows (Eigen 1976; Eigen and Schuster 1979; 1982; Eigen and Winkler-Oswatitsch 1992):

1. The individuals among which selection is to take place must be self-reproducing entities; they are replicators.
2. Self-reproduction is subject to errors. This means that some replicators come into existence not as a result of true copying of an identical parent, but as a consequence of inaccurate copying of one that is closely related.
3. Self-replication must take place far away from chemical equilibrium.

According to Eigen these three principles represent the *necessary internal*[4] conditions for the beginning of the process of Darwinian evolution and selection. Together they have the character of "a clear 'if-then' principle; that is, a principle according to which defined initial situations lead to deducible behavior patterns"

(Eigen and Winkler-Oswatitsch 1992, 18). With respect to that necessary internal conditions it holds that "[a]nything which *can* form under [these] appropriate conditions *will* do so" (Eigen 1971, 470). The process of evolution and selection is occurring "inevitably wherever and whenever those criteria are fulfilled" (Eigen and Winkler-Oswatitsch 1981c, 282).

Those principles provided Eigen with the starting points for the understanding of causal relations at the level of biological macromolecules in a loop-like manner, and for a mathematical treatment of their evolution. The latter, in its turn, enabled him to view the concept of hypercycle as central for the understanding of the development of life on Earth.

Eigen suggests a three-stage hierarchy of reaction cycles, where the latter is a type of reaction in which the products undergo further transformations, so that any product formed is of the same type as the reactant of a preceding step (Eigen and Schuster 1979, 3–6). Stage one can be represented as follows:

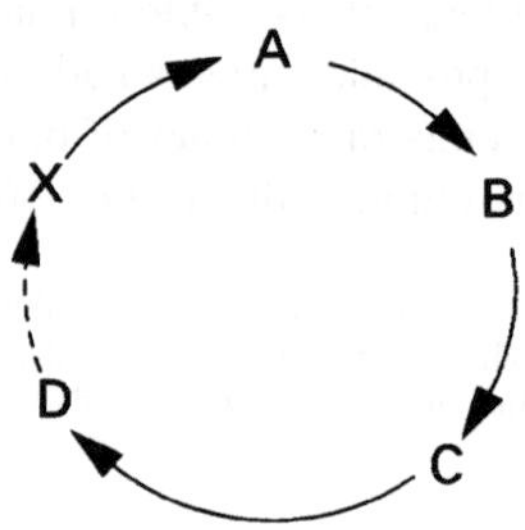

Figure 5.

Here *A* catalyses the creation of *B*, *B* that of *C*, *etc.*, so that finally *X* acts back on the creation of *A* (Eigen-Winkler 1981, 220–221). The whole cycle is a catalyst and can be designated by the symbol *E*. The symbols *A*, *B*, . . ., *X* represent intermediates which turn certain substrates, *S*, into products, *P*.

Stage two is

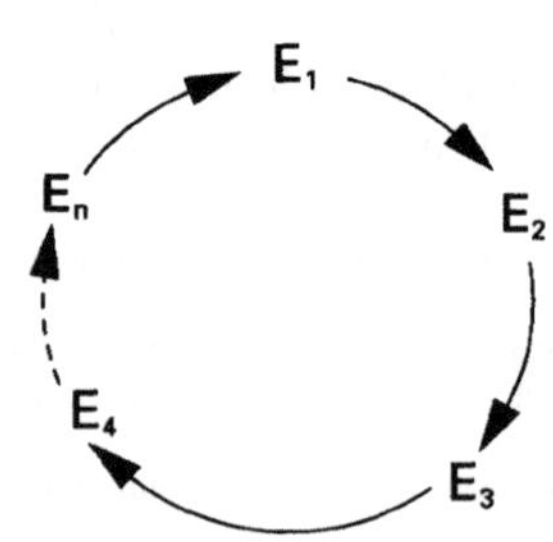

Figure 6.

Here the constituents of the cycle $E_1 \rightarrow E_n$ are themselves catalysts, so that each intermediate E_i (formed from some substance *S*) is a catalyst for the formation of E_{i+1}.

This process for a whole class of information-carrying entities I_i can be formally represented as $S \xrightarrow{I} I$, and the whole cycle can be designated by the letter I. Finally, stage three is

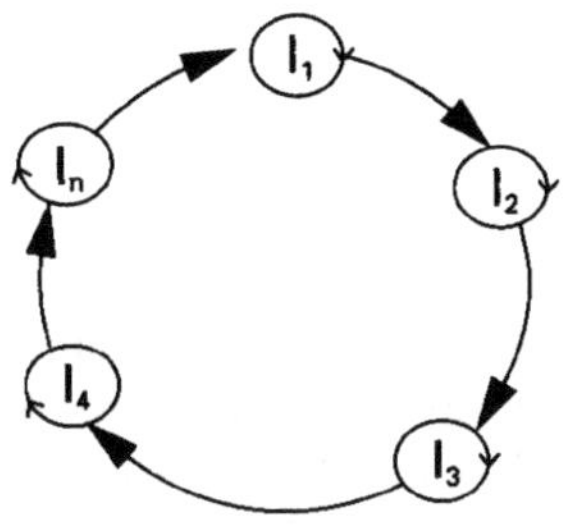

Figure 7.

The entity here consists of self-instructive units ; it is a hypercycle. I_i instructs not only its own reproduction but also provides catalytic support for the reproduction of I_{i+1}. Here we have a cyclical linkage among self-reproductive individuals; a cycle of cycles. Their interconnection has the character of a self-enhancing cycle.

For a mathematical treatment of evolution and selection, which Eigen views as a necessary result of the above given three conditions, he introduces the following system of differential equations conforming them (Σ stands here for a sum so that $i \neq k$ holds):

$$dx_i/dt = (A_iQ_i - D_i)x_i + \Sigma w_{ik}x_k + \Phi_i \qquad (9.1)$$

Here the index i stands for all distinguishable self-reproductive individuals, so that $i \in <1, N>$, while x_i stands for their concentration, $A_iQ_ix_i$ expresses the spontaneous formation due to the process of metabolism, while D_ix_i stands for decomposition in this process. Φ_i stands for the outflux of the reaction products, Q_i is the quality expressing that part of the reproduction of the individual unit i which results in an exact copy of i. The fact that the reproduction of i leads to error copies, which are still closely related to i, is expressed as $A_i(1 - Q_i)x_i$. Due to the mutation process in the reproduction of these "relatives," x_i also receives a contribution from this process; this is expressed as $\Sigma w_{ik}x_k$.

If each species involved contributes to the total flow Φ_t in proportion to its presence and Σx_k stands for the overall population density, it holds that $\Phi_i = \Phi_t x_i/\Sigma x_k$. Under the supposition that $\Sigma x_k = const = c_n$ (constant overall population density), the total flow has to be regulated so as to compensate for the overall excess production. For $E_i = A_i - D_i$ as the excess production it holds that $\Phi_t = \Sigma A_kx_k - \Sigma D_kx_k = -\Sigma E_kx_k$. (9.1) then changes into

$$dx_i/dt = (W_{ii} - E(t))x_i + \Sigma w_{ik}x_k \qquad (9.2)$$

where $W_{ii} = A_iQ_i - D_i$ is the intrinsic selective value and $E(t) = \Sigma E_kx_k/\Sigma x_k$ is the average excess productivity.

Because the total number of the population is conserved, the closely related species cooperate, while the more distant enter into mutual competition. What initially was presupposed to be one species of individuals is, according to Eigen, in fact created by closely related individuals; they do not create a species but a *quasi-species*. So instead of supposing the existence of N species, one has a set of N quasi-species, where their population variables y_i are linear combinations of the variables x_i , so that $\Sigma x_k = \Sigma y_k$ holds and (9.2) changes into

$$dy_i/dt = (\lambda_i - E(t))y_i \qquad (9.3)$$

where λ_i are the eigen-values of the dynamic system. From (9.3) Eigen draws the following conclusions. $E(t)$ as the average excess productivity represents a threshold. Any quasi-species characterized by y_i and λ_i, for which $\lambda_i < E(t)$ holds, dies out. Only if it holds that $\lambda_i > E(t)$ the quasi-species will grow. On the supposition that $E(t)$ (as a function of time) will grow the more the system prefers quasi-species with large λ_i, until it will finally reach a steady state (i.e., that holds $E(t) \rightarrow \lambda_{max}$), the average excess productivity will reach a maximum. This in turn means that only one of all the quasi-species, namely, that which is assigned to λ_{max} will have survived; all others will have died out in the process of (Darwinian) selection and evolution. Accordingly (Eigen and Schuster 1979, 10)

Darwinian selection and evolution can . . . be characterized by an extremum principle. It defines a category of behavior of self-replicative entities under selection constraints.

The quasi-species represents (see, e.g., Eigen and McCaskill and Schuster 1988) a mutant distribution in which the standard (master) copies quantitatively dominate with respect to any other nonstandard type, even if, with respect to the total population of the quasi-species, those copies represent only a small fraction of it.

By mathematically treating the concept of a quasi-species even further, Eigen theoretically reconstructed the next stage in the evolution of life. If σ_m stands for the superiority parameter rating the reproductive performance of the master copy relative to its mutants, and $1 - q_m$ is the average error-rate per symbol in it, it is possible to derive a relation for the maximum information content of a quasi-species. The relation is $\nu_{max} = ln\sigma_m/(1 - q_m)$. It states that (Eigen and Schuster 1979, 15)

[t]he number of molecular symbols of a self-reproducible unit is restricted, the limit being inversely proportional to the average error rate per symbol: $(1 - q_m)$.

This means that in the evolution of life on Earth there was a stage when the growth of the information content of a quasi-species could not be realized any more inside the self-replicative unit itself. Given its error rate, its length – the number of its symbols – was limited. Any increase of the latter would cause the "evaporation" (due to the error-rate) of its genetic information, i.e., the destruction of its order. What is required for the understanding of the further growth of the information content of living entities is a new principle of macromolecular organization.

In order to discover this new type of organization, Eigen again applied game theory together with a computer simulation of error-generation in a set of words which were cyclically generated by a computer program.[5] Eigen suggested three possible alternatives (Eigen and Schuster 1979, 25–26): the words are unrelated and plainly coexist; the words are functionally linked so that a word provides "catalytic" help for the reproduction of the next word; finally the couplings among the words form a closed loop. The results of the computer generated game show that in the first case the whole sentence is unstable. Even if each word is assigned the same selective value, only one word survives in the competition. In the second case the last word always "outgrows" all the others, and for the case that the couplings between them are proportional to the population number of the catalyst, only the last word survives. Finally, in the third case the information content of the whole sentence is stable, and it is possible to get beyond the threshold-value ν_{max} for each individual catalyst involved.

The importance of the concept of hypercycle lies, *firstly*, in the fact that, together with the above derived extremum principle of election, it provides the basis for the explanation of the second problem (unity) of evolution. By further developing the above-given mathematical treatment (Eigen and Schuster 1979, 28–31), Eigen proves that under the constraint of constant organization given above and under the supposition that the self-replicative entities are characterized by a hyperbolic rate of growth, Darwinian selection has a sharp, once-for-ever character (Eigen and Schuster 1979, 30–31; Eigen and Winkler 1981, 230–235). From the competing self-reproducing systems, one is selected; all others die out, and no newly-created mutant can challenge an established population selected in such a way. The fact that all now-existing living organisms use the same coding system and the same protein machinery seems to be the result of such a type of Darwinian selection.[6]

The importance of the concept of hypercycle lies, *secondly*, in the fact that Eigen views this form of macromolecular organization as a *necessary* result of the existence of entities, fulfilling the above given three conditions of evolution and selection, if they mutually interact. Initially these interactions could be weak and accidental, but once given they will "lead to quite new *consequences* . . . a hypercyclic organization will inevitably emerge whenever such interactions arise . . . Its emergence was inevitable, whenever the conditions laid down by nature allowed it" (Eigen and Schuster 1979, 75, 87).

9.3 SUBSTANCE-SUBJECT vs. "BACKWARDS" CAUSATION

Mellor's approach to the idea of causal loops[7] rests on the supposition that the individual event causing the event e^* is the same as the individual event which is caused by the event e^*; therefore one can assign just one symbol e. But having in mind that a singular event is an entity which exists in a certain time-span, one should rather suppose that the occurrence of the singular event-effect e^* means that the singular event-cause e has already disappeared, passed out of existence. So when the event e^* causes another event – even if the latter could be somehow similar to e – still it is different from the singular event e. One should, therefore, assign to these "similar" but still different singular events two different symbols: e_1 and e_2. This would solve the paradoxical result of Mellor's reflections on causal loops, namely, that one and the same event should have two different probabilities of occurrence.

H. Reichenbach's approach to causal loops aims exactly in this direction. In (1957) based on his (1928), he presents, as compared to (1956), a more differentiated approach to causal loops. He starts with the idea of determining time-relations on the basis of causal relations. For two events E_1 and E_2 so that "E_1 is the cause of E_2, then a small variation (mark) in E_1 is associated with the small variation in E_2, whereas small variations in E_2 are not associated with variations in E_1" (1957, 135). If E^* stands for a marked event, then one can observe only the cases E_1E_2, $E_1{}^*E_2{}^*$, and $E_1E_2{}^*$, but never the case $E_1{}^*E_2$. "That event which appears in the unobserved combination without an asterisk, namely E_2, is called the effect and furthermore the *temporally later event*" (1957, 137). Reichenbach then suggests the case symbolized above by Figure 4, so that "there is a causal chain from E_1 to E_2 such that E_2 is an effect of E_1, while there is also a causal chain that makes E_1 an effect of E_2" (1957, 139). According to him (1957, 139)

> [t]he combination would result in a causal chain that returns to its origin. In order to exclude a contradiction we must make the assumption that there are *no closed causal chains*.

Finally, Reichenbach suggests the mechanism of an electric bell by the following scheme, which seemingly suggests that it functions on the basis of loop-like causal relations.

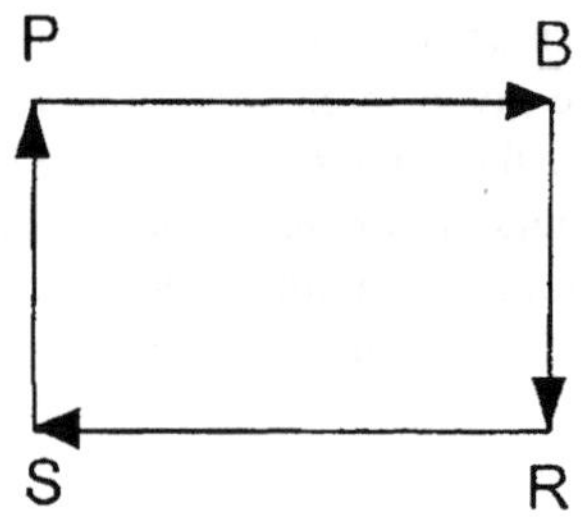

Figure 8.

and states (1957, 139–140)

> [t]he pulling (*P*) of the lever causes a break (*B*) in the current, which in turn causes the return (*R*) of the lever; this switches on (*S*) the current which finally pulls the lever. It appears as though this chain could be diagrammed [by] . . . a closed curve. The mistake in this argument, however, is easily seen. The individual pulls of the lever are different events, i.e., although they are of the same kind, they are not identical events. Consequently, the chain should be diagrammed as . . . an open chain

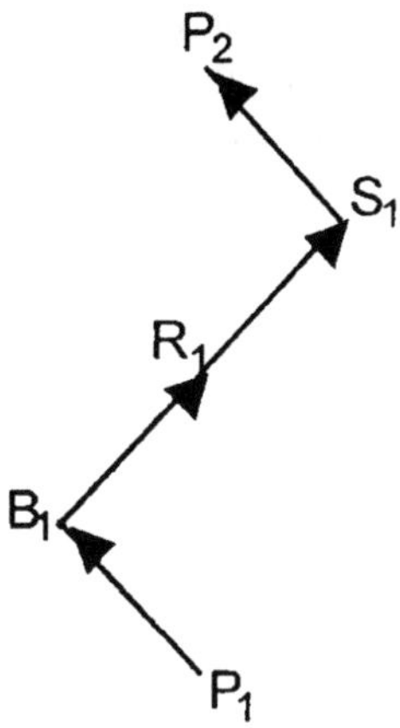

Figure 9.

Our principle of the mark forces us to this conception. If we make some change in B_1 by short-circuiting the current and thus preventing the return of the lever, then P_2 will no longer occur. However such interference does not change P_1.

According to Reichenbach, the moral of such an approach to causal loops, which goes straight against Mellor's understanding of them, as well as against any attempt to deal with the issue of causal loops by means of a singular-event-approach, is as follows (1957, 140):

[n]ow it is clear what we mean by a closed chain, namely a chain that returns to *identically the same event*, not to one of the same kind.

Even if one accepts Reichenbach's view that there can exist loop-like causal relations between different kinds of events, one still one the same problem as reconstructed by Mellor. By substituting into the probability relations symbols representing kinds (classes, types) of events, one ends up with the result that the probability of a kind of event e – if it is a cause – is different from its probability when it is an effect. So again it is possible to claim that the idea of causal loops is just empty speculation.

For the solution of the issue of causal loops one has to realize that the approach to loop-like causal relations based on the probability calculus targets only (singular or kinds of) events which are the suppositions or results of causal processes. What is left out are the very causal processes. The crucial point for the understanding of loop-like causal relations is the that the processes starting from a kind of event, state, "fact," *etc.*, are different from those which lead to their production. Expressed otherwise; *the mechanism of the process triggered by the cause is different from the mechanism of the process which (re)produces it as an effect.* This fact is readily seen in Reichenbach's bell example. The pulling of the lever generates the break of the current, but then other processes, different (from the former) step in (the break in the current generates the return of the lever; the return of the lever switches the current), so that finally the lever is pulled again. The same holds for Eigen's realistic hypercycle with translation. He represents it as follows (Eigen and Schuster 1979, 5, 50):

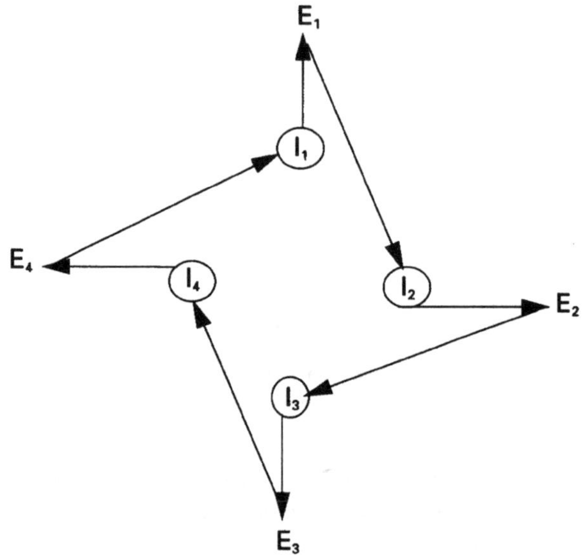

Figure 10.

For the catalyst I_1 it holds, for example, that it instructs the production of the intermediate (polypeptide) E_1, but then between E_1 – as the effect of the cause I_1 – and I_1 as the result-effect, other, different causal processes come in.

Suppose now[8] that a traveler in a miniaturized "space-ship" travels from B through R and S back to B in a device represented by Figure 8 or from I_1 through E_1, I_2, E_2, I_3, E_3, I_4 and E_4 back to I_1. What can the traveler observe? There are two options. One is that during his trip there are no changes in the constituents of the cycle, e.g., marking – like in Figure 9 – of the process of functioning of the bell does not take place, or there are no mutations of constituents of the hypercycle. Then the traveler will be able to perceive a direction of time only if he uses a measure external to that cyclical process, say the aging of his own body; without such a device he would perceive the processes inside the bell or the hypercycle only as an "eternal return." He would not of course need such a device in the case when the constituents of the reproduction change by themselves, say, by mutations, as in the case of Eigen's hypercycle, or by the process of accumulation as explicated in Marx's reconstruction of the reproduction of the capitalist economy.

While Reichenbach in his bell example supposes that the process of its functioning is *marked from the outside, I understand the process of non-simple reproduction*, reconstructed in Chapter 5, *as a process which marks itself*. This conclusion enables us to look from another point of view at the issue, discussed in Chapter 7, of the possibility vs. impossibility of grounding the direction of time in a system. The idea that the direction of time in a system has its ground in the non-simplicity of its own reproduction appears now as a *causal understanding of the direction of time*.

This understanding also shows from another point of view why Prigogine's attempts to arrive at a new understanding of time fail. He supposes that the environment (the outside) controls and regulates the parameters generating the bifurcations of the system's trajectories. *The environment has to mark the Brusselator in order to generate the becoming of an organized, complex structure, and, therefore, also generate the direction of time*. In Eigen's approach, in comparison to Prigogine's, two new ideas come in. The first is the idea of a non-simple reproduction introduced by two of the principles of Darwinian selection and evolution: self-reproductivity and mutagenity. The second idea is that the presuppositions-inputs and results-outputs of a self-replicating unit are cyclically interconnected by means of other self-replicative units inside the system. These two ideas in turn enable Eigen to understand why a living system can escape the control from the outside, why it genetically drifts, and why Darwinian selection takes place, even if there is no "external selector" (Eigen and Schuster 1979, 7).

What are further characteristics of an object capable of non-simple reproduction? All its determinations, subparts, and particulars are cyclically connected, so that they *mutually produce, constitute each other*. The object given as their processual unity is, therefore, an *active entity*. Because of its own self-reproduction the object is something *constant, regular, enduring*; because of the non-simplicity of this reproduction it is also something *changing, having directionality* and finally, as shown in Chapter 5, *historical*. The history of this object is its own history. The time between its historical becoming and historical decline is its own *internal time*. *Such an object is not a passive substratum shaped from the outside, but a substance which shapes itself, which is the source, subject, starting point and end of all its activities and movements on its own ground: functioning, development, and decline*. Such an object can properly be labeled *substance-subject*. In Chapter 11 I will compare Hegel's[9] approach to this category with that of Marx.

Now it is possible to evaluate Mellor's and Mackie's ideas on "backwards" causation. It is a plain fact that there exist cyclical causal relations in the real world. Loop-like relations are still present in the machinery of the living cell as it exists today. In it the whole process of copying, reading, and translation "is ultimately a huge autocatalytic feedback loop . . . and the entire machinery . . . can therefore be said to 'produce itself'" (Eigen and Winkler-Oswatitsch 1992, 61). Loop-like causal relations are also given in the infection cycle of virus upon which viral diseases are based.[10] The scheme of the cycle of plus strand virus is as follows (Eigen and Schuster 1979, 5; Eigen and Winkler-Oswatitsch 1992, 34).

which I view as the most important and which will serve as a starting point for stating my views on universals.

1) Realism is a doctrine according to which in the real world there are not only particulars.
2) One has to admit universals.
3) Particulars cannot exist without properties (*The Principle of the Rejection of Bare Particulars*). Neither does there exist a bare universal (a universal without a particular or particulars).
4) The union of particulars and universals is given in a state of affairs, which is defined as a particular's having a property or more particulars being related by a relation. The mutual dependence of particulars and universals means that neither can exist in independence of states of affairs.
5) Properties are monadic universals, relations are polyadic universals.
6) The world consists only of particulars having properties and relations (monadic and polyadic universals), where these universals themselves have relations.
7) A property must be a property of some real particular; a relation must hold between real particulars (*The Principle of Instantiation*). A universal need not be instantiated now; what is real is not confined to the present.
8) If it can be proved *a priori* that a thing falls under a certain universal, then there is no such universal. A property or a relation of a particular endows it with a specific causal power; what its characteristics are can be find out only *a posteriori* (*The Principle of Method*).
9) Universals are the repeatable features of the world.
10) A particular is never related to itself; any doctrine of *causa sui* is rejected.
11) Universals as relations and properties of particulars are first-order universals and these particulars are first-order particulars (absolute particulars). Higher-order universals exist but they are formal universals (*Formalist Principle*); they make no special contribution to the unique set of powers bestowed by the first-order universals.
12) Second-order properties have no properties, but second order relations may fall under universals (they could have properties).
13) Laws of nature (causal and nomic connections) are irreducible second-order relations between first-order universals, genuine relations between universals.
14) If *F* and *G* stand for certain properties, then the fact that it is a law that *F*s are *G*s can be expressed as "Being an *F* necessitates being an *G*" or "Something's being *F* necessitates that same something's being *G*, in virtue of the universal *F* and *G*." This can is expressed as *N(F, G)*.
15) If *F* and *G* are related by a dyadic relation of necessitation, then they have this relation omnitemporally and omnispatially. There was no time where they gained it and no time where they will lose it. They either have it or lack it.
16) Since nothing is genuinely related to itself (point 10, above) the relation of necessitation is necessary irreflexive. There can be no law of nature of the form *N(F, F)*.
17) The relation of necessitation is a *contingent* relation; if *F* and *G* are related by *N*, it is possible that they should not have this relation.

With this summary of Armstrong's view on universals, let me now analyze certain aspects of it.

Firstly, in Armstrong's approach a particular exists as a particular independently of other particulars. So, e.g., he claims for the case of the father-child relation (1978, Vol. II, 85) that *f is the father of c* cannot be taken as a relational *property* (a universal) *given by the relation of the particulars*, but has to be understood as a universal belonging to the particular *f*: being a father of a child.

Secondly, Armstrong claims that universals are the repeatable features of the real world, but no ontological explanation of this repetition is given.

Thirdly, when one looks thoroughly at Armstrong's hierarchical construction of the world, he finds out that while there is a mutual dependence between first-order (absolute) particulars and first-order universals (The Principle of Instantiation *and* The Principle of the Rejection of the Bare Particular), there is no such mutual interdependence between the dyadic relation of necessitation *N* and the first order universals *F* and *G*. Not only does *N* add nothing to the powers bestowed by *F* and *G*, but also a concrete *N*, given a certain *F*, does not generate a certain *G*; *N* relates them, so to say, *post festum*.

Because *N* adds nothing to *F* and *G*, and *F* can exist outside the *N*-relation to *G* and *G* can exist outside the *N*-relation to *F*, Armstrong views the very *N*-relation as an *ahistorical relation*. In his view there is no such mutual interdependence between *F* and *G*, i.e., no such *N*-relation which becomes (originates), develops, and finally declines (disappears); so there is no mutual generation nor mutual (re)production of *F* and *G*.

Fifthly, Armstrong rejects the idea that a particular can be related to itself, and, as a conclusion, he rejects any doctrine of *causa sui*: His proof is as follows.

1) Let "x", "y" and "z" are suitably bound variables and it is allowed to make unrestricted substitutions within the range of these variables.
2a) If xRy, then yRx; and
2b) if xRy and yRz, then xRz.
3) If xRy is given, by 1) yRx is given as well. So we can assert xRy and yRx.
4) By substituting *x* for *z* and if the antecedent of 2b) is instantiated, the conclusion is that if xRy, then xRx. Expressed otherwise: "[i]f x has R to anything, then it has it to itself" (1978, Vol. II, p. 93). And he concludes that "we can allow the deduction but deny that 'having R to x' applies to x in virtue of a relation in the world x has to itself. It would be a predicate which applies, but in not in virtue of any universal" (1978, Vol. II, p.93).

Item 2a) appears to be the central premise in this proof. But my reconstruction of the concept of causal loops shows that 2a) has to be restated into: "If xRy, then essentially necessarily yR′x" and where R′ ≠ R and, as I will show below, both R and R′ stand for a certain type of necessity. But then 2b) does not hold and neither does the whole proof.

Sixthly, Armstrong's rejection of the idea of *causa sui* surfaces in his critique of C. Swoyer's approach to laws of nature. Swoyer amends Armstrong's approach to laws of nature in such a way that the relation of necessitation between universals is not contingent but a necessary one. He, therefore, views properties of particulars as having essential features, where these are given by the relations (so-called relations of nomic implication) of these properties to other properties, and where these relations set up a circle. Because Armstrong rejects the idea of *causa sui*, he criticizes this idea of a circular relation between essential features as follows (1983, 160):

But what of the circle as a whole? What is essential to it? Nothing, presumably. It is just itself. But if it can be just itself, why not a 'circle' of one, a single property which has no essential feature, but is instead itself?

Of course, if one understands the essence as a mutually differentiated relation of entities which produce and reproduce each other, then the answer to Armstrong's question is: *It is its own essence; it is the essence of itself.*

My approach to the concepts of essence of a natural kind and of causal loops yields the following approach to universals.

A universal of a natural kind is a unity of mutually differentiated entities (things, properties, relations) which set up the causal basis for the particulars and individuals belonging to this natural kind. What unifies these entities into a universal is that they set up an operationally closed circle of mutual relations; and by means of it they constitute each other. They exist as these entities only in this universal. Only through their mutual relations are they able to produce and reproduce the very inherent (essential) conditions of the existence of this universal, i.e., of their common being, and, therefore, also produce and (re)produce their own existence. Because this universal has the character of the substance-subject, I label such a universal a substance-universal. This category partially draws upon Hegel's category of the concrete universal; I will analyze it in Chapter 11. The substance-universal is a form of reproduction of those conditions in which the common being of these entities can exist, function, develop, and finally decline. The substance-universal is, therefore, also a repeatable feature of the real (natural or social) world. So, while Armstrong cannot ground this repeatability, I ground it in the very internal structure of the universal, in its capability to reproduce the very conditions of its own existence. Such a universal occurs everywhere and only where the conditions of its existence occur. Everywhere these conditions occur, this type of mutual relation of entities inevitably comes into being; it has the character of a *de re* logical necessity. So the substance-universal is also a form of a real and historical necessity; it has the character of a real, objective (natural or social) law. And it is completely irrelevant if those conditions occur in the whole universe or only in a part of it. In such a way I am able to ground the contention of the necessitarian approach to scientific laws and real (natural or social) laws, that "it is counterintuitive to make the lawlikeness of a universal dependent on the number and distribution of instances" (Suchting 1969, 351).

Now it is possible to improve on Armstrong's approach to universals. The *N*-relation of necessitation between universals *F* and *G* bestows upon the particulars instantiating them the fundamental power to have *F* and to have *G*. It bestows upon them their very existence as particulars. *F* and *G*, as well as the particulars instantiating them, can exist only through the *N*-relation, one particular by means of the other particular. So there is not only a mutual dependence between a first-order (absolute) particular and a first-order universal, as claimed by Armstrong, but also a *mutual, bi-directional dependence between second-order universals and first-order universals*. To exemplify this let me pose the relation of the necessary conditions of Darwinian selection and evolution to the process itself: this very Darwinian selection and evolution. The necessary conditions were historically a result of processes completely unrelated to the process of evolution and selection. But once they *caused* the latter, they themselves became the *necessary result,* or *effect* of it. The three necessary conditions of Darwinian selection and evolution are reproduced by the very process of selection and evolution as its result. According to Eigen, these conditions are not only permanently at work at the level of the first polynucleotides with the restricted length ν_{max}, but also at the level of hypercyclic organization. The evolution of the hypercycle is based on mutations of its constituents and on cyclical relations of its constituents by means of which they enhance each other (Eigen and Schuster 1979, 72–75). This dependence can be symbolically expressed by means of the operator for essential necessity N_e. Again, as in the case of the sentential connective from Chapter 5, symbolically represented as "$\underset{r}{\rightarrow}$", I develop here neither a syntactical nor a semantical analysis of this operator; the latter will, it seems to me, depend on the former. Let N_{e1}(F,G) express that if F is given, then G is essentially necessarily produced. Let $N_{e2}(G,F)$ express that whenever G is given F is essentially necessary produced. The fact that the N_{e1}-relation and the N_{e2}-relation together, through their mutual reproduction, set up a system which produces and reproduces itself, i.e. set up a *causa sui* system, I express as $N_{e3}(N_{e1}(F,G),N_{e2}(G,F))$. N_{e3} stands here for the essential necessity of the occurrence of $N_{e2}(G,F)$ whenever $N_{e1}(F,G)$ occurs and *vice versa*. N_{e3} is a third-order universal which bestows upon the second-order universals N_{e1} and N_{e2} their very existence, i.e., it bestows a *specific power*. Because there is a bi-directional relation between second-order universals and first-order universals, there also exists a mutual relation between the second-order universal and the particular instantiating that first-order universal; and, therefore, there also exists a bi-directional relation between the absolute particulars instantiating those first-order universals. The particular and the universal presuppose each other. The universal can exist only due to the mutual relations of the particulars, but the very particular is what it is only due to the universal; only through it can it exist and function. The particular reproduces the very conditions of its own existence only by means of this universal.

My claim that a universal, understood as the relation between mutually differentiated entities, has the character of a *de re* logical necessity faces a problem which can be labeled as *the problem of vacuous (uninstantianted) laws of nature*. It haunted Swoyer (1982) and served Mellor (1980) as an argument to reject Armstrong's as well as Tooley's (1977) and Dretske's (1977) approaches to laws of nature. The problem can be stated as follows. Let us take the scientific law relating the vapor pressure, P, of water to its temperature, t. The respective equation is $lnP = A - B/t$, where A and B stand for certain constants. The problem is that there are certain values of t which water cannot attain because it decomposes before it reaches them. But this contradicts my claim that $N_e(F,G)$ holds *de re* logically necessary. If $N_e(F,G)$ holds *de re* by logical necessity, then that equation should hold in all possible worlds where water exists, but, as mentioned earlier, there are certain possible worlds where, due to the given temperature, water could not exist, and where, therefore, the equation is useless.

Mellor pushes this argument even further. He claims that even Armstrong's and Tooley's view on laws of nature, namely that $N(F,G)$ is only a non-logical (contingent) necessary relation, is wrong because it cannot cope with the problem of vacuous laws. He states "[t]o do its job, N has not only to make Gs out of the actual Fs, it has . . . to be such that anything were it F, would be G" (Mellor 1980, 120). So $N(F,G)$ can be stated as

(3) If x were F it would be G.

Seemingly the following case corresponds to (3). Let us have an individual, denoted by the individual constant "a" which at time t has the inertial mass m. Then all conditionals of the following form hold:

(4) If a were subject at t to a force of f newtons, it would then accelerate in the direction of the force at f/m metres/second^2.

One views all such conditionals as true "and a fact is needed to make them so; and the requisite fact is that *a* has mass *m* at time *t*" (Mellor 1980, 120–121). The condition of *a*'s having mass *m* at time *t* is, in the words of P. Tooley, the truth-maker for conditionals of type (4). But here, according to Mellor, the crucial difference between (3) and (4) surfaces. The truth of (4) entails that *a* exists, for without *a* – as the reference of "*a*" – (4) would lack any truth value. But, he claims, the truth of (3) does not entail that the universal *F* exists, that *F* is instantiated, that some *F*s exist. If one follows Armstrong and claims (as he himself does) that the *Principle of Instantiation* has to hold, there will be no fact *N(F,G)* "to make (3) true of everything; and the problem of accounting for vacuous laws will remain unsolved" (1980, 121).

How to cope with the problem of vacuous laws? Primarily, in my view, one has to get rid of the illusion that it pertains to the very laws of nature as something real in the world. This illusion has its roots in the ambiguous usage of the concept "natural law." That this is so can be readily seen from Swoyer's approach in (1982). Even if he strictly distinguishes between *natural laws*, as something metaphysical, ontical (objective facts) and *statements* of natural laws. as something epistemical, still he claims "[t]hat laws of nature play a vital role in explanation, and inductive inference is far clearer than the nature of the laws themselves" (1982, 203). So the issue of vacuous laws covers in fact two problems. The first is the problem of the givenness, the existence of real laws, relating certain entities (processes, things, relations, etc.), in those parts of the universe where these entities are not given. In my view, in those parts of the universe where these entities are not given, the laws of nature relating them are not given either. There are no truth-makers for scientific laws which special sciences could formulate and apply to those parts of the universe.

The second problem pertains to scientific laws and contains two issues: one is of a semantical nature, and the other of an epistemic, cognitive nature. The semantic issue pertains to the truth/falsity/lack-of-truth-value of statements containing universal names, which in some possible worlds lack extension. Mellor in his example about the conditional form of Newton's second law of dynamics implicitly draws upon Frege's semantics. In it, a statement lacks reference (truth-value) if a proper name contained in it lacks reference. So, only from Frege's semantics for proper names, plus the fact that (4) is true, does it follow, that the reference of the proper name "*a*" exists. But Mellor's claim that from the truth of (3) it does not follow that the universal *F* exists is based on the absence of a premise which was crucial in his argument that the reference of "*a*" exists. It lacks a semantics for the universal name "*F*" whose extension is the universal *F*. The issue of the truth/falsity/lack-of-truth value of statements of the form "*N(F,G)*", in the case when "*F*" lacks extension can be solved only in the framework of a semantics for universal names appearing in scientific theories.

The second issue pertains to the epistemic problem of the cognitive limitations of certain types of scientific law. To make this point understandable, let me take Mellor's example of the law for the vapor pressure. It is formulated on the supposition that water exists. It does not, and that is its inherent epistemic limitation in it, relate the very existence of water to temperature. The epistemic lesson which can be drawn from this example of Mellor is the same as the lesson drawn in subchapter 2.6 from

Mackie's purification of the laws of mechanics. A certain concrete scientific law need not recognize all the aspects and characteristics of entities which are grasped by it. It is not an all-or-nothing issue. Special sciences formulate scientific laws which differ from each other in their respective epistemic depth. And philosophy of science should not take less developed concrete scientific laws, scientific laws with a lesser cognitive depth (e.g. scientific empirical laws), as a proof for its claim that special sciences characteristically cannot formulate more developed scientific laws, laws with a greater cognitive depth which grasp hitherto uncomprehended aspects and characteristics, e.g., the *de re* logically necessary relations between universals.

Such a differentiated view of special sciences enables me also to evaluate Armstrong's realistic approach to universals. In my view it is sufficient for special sciences which do not analyze the reproduction processes and do not reconstruct the origin of their objects. Such a type of science is, e.g., classical mechanics. It operates only with the idea of the unidirectional action of forces, and it investigates only their effects, not their origin. But Armstrong's approach is not sufficient for special sciences which deal with issues like reproduction, origin, and becoming, issues which come to the fore, for example, in the case of molecular biology or political economy.

PART 4

THE EPISTEMIC BACKGROUND

In life categories are *used*; . . . they *serve* in the creation and exchange of ideas involved in the spiritual activity (geistiger Betrieb) on a living content . . . Such a use of the categories, which above was called natural logic, is unconscious . . . To purify (zu reinigen) these categories, which as instincts (Triebe) are only instinctively active, that is the loftier business of logic (Hegel 1923, T. 1, 13–16; 1969, 34–37).

Metaphysics is nothing but the range of universal thought-determinations, and is as it were the diamond-net into which we bring everything to make it intelligible. Every culturated consciousness (gebildetes Bewußtsein) has its metaphysics, its instinctive way of thinking. This is the absolute power within us, and we shall only master it if we make it the object of our knowledge . . . All culturation (Bildung) reduces itself to a difference of categories. All revolutions, whether in the sciences or world history, occur merely because spirit now for its self-understanding and self-perception . . . has changed its categories . . . (Hegel 1842, §246, Zusatz, 19; 1970a, 11; 1970b, Vol. 1, 202).

CHAPTER 10

EPISTEMOLOGY AND LOGIC

The aim of this chapter is to provide a background for my earlier analysis of scientific laws and of scientific explanation. Till now I was operating in the framework of the philosophy of science. Now I will show that the issue of scientific laws and of scientific explanation is but a moment of a broader epistemic setting, requiring analysis in the framework of the theory of knowledge (epistemology). Frege's understanding of the concept of sense (*Sinn*) suggests itself as a convenient starting point here. By analyzing it I show that it has an epistemic character, and that epistemic character (along with its development) is the ultimate subject matter of philosophy. Philosophy, understood here a synthesis of a newly comprehended ontology, epistemology, and logic, should analyze thinking by means of philosophical categories and set as its target the creation of a system of philosophical categories.

10.1 FREGEAN SENSE, THE REALM OF SENSE AND ANALYTIC PHILOSOPHY

Frege in his *Über Sinn und Bedeutung* introduces the term "sense" as it relates to proper names by speaking about "valuable extension of our knowledge" (*wertvolle Erweiterung unserer Erkenntnis*), about the "knowledge of the reference" (*Erkenntnis der Bedeutung*), and about "cognitive value" (*Erkenntniswert*) (1892, 56, 58, 78). In the case of a statement, sense represents what we know when we know that it is either true or false, or what we know that we have to find out to determine its truth-value. Therefore, Fregean sense is a cognitive, epistemic concept. Frege also claims that it is the objective content of our thinking, "which may be the common property of many" (1892, 59). So, I understand sense also as the being of the (cognized) thing in us, the collectively owned informative representative of the real object, the thought-object *(Gedankending)*.[1]

The issue which has to be cleared up can be formulated in the following question: Where does Fregean sense reside? By claiming that sense is a collectively owned informative representation of the real object, I have departed in an important aspect from its understanding by Frege. While I understand sense as knowledge (fixed in a language) about an object, as something epistemic, cognitive, Frege in fact understands the sense of a name not as something epistemic, but only as something close to the epistemic; or more exactly: as something which is definitely closer to the epistemic than the reference of this name. My claim was founded on the idea that sense is collectively owned. But for Frege it can, but need not be, collectively owned. According to him, sense does not require human beings at all for its very existence. To understand this, one has to bear in mind that Frege fought a life-time-battle against psychologism. While analyzing language, he distinguished not only between sense and the reference of a name, but also mental (psychic) states generated by the name in the minds of the individual name-users. He sums this states up in the word "idea" (1918a, 14), where ideas are exemplified by a name's coloring (*Färbung*) and tone (*Beleuchtung*) (1906b, 193), by creations of imagination, by sensations, feelings, and moods (1918, 13–14). All these products of an individual mind are, according to Frege, the subject-mattter of psychology and not of logic. On the other hand, senses (thoughts) are *eternal immutable entities* which can be grasped by human beings. But the existence of thoughts (senses) is not mind-dependent; mind stands in relation to them as to something external (1897, 133), and "[o]ur act of judgment can in no way alter the make-up of a thought . . . A true thought cannot be affected by our act of judgment" (1918b, 37). Even if there were no human beings at all, the realm of sense would still exist; the latter does not require a bearer, an owner (1918a, 17–18). Frege even introduced a realm of sense, labeling it "third realm" (*das dritte Reich*), thus distinguishing it from the world of ideas whose owner, or bearer, is the individual's mind, and the outer world (world of reference[2]) (1918a, 13–15). The third world exists prior to and independently of human thinking; the latter does not create the former, "but only comes to stand in a certain relation to what already existed" (1918a, 18). "In thinking we do not produce thoughts, we grasp them," (1918a, 25) or, in other words, "[w]hat is grasped, taken hold of, is already there and all we do is take possession of it" (1897, 137). The process of grasping thought has a mental (*seelischen*) character (1897, 145), it goes on in the world of ideas, and is not the subject matter of logic, but of psychology (1897, 145): "[b]oth grasping a thought and making a judgment . . . are to be assigned to psychology" (1919, 253).

The importance of the above ideas of Frege for the future development of philosophy, especially for its analytic branch, lies in the fact that it had a decisive impact on, as well as an anticipational dimension with respect to, that development. Dummett proved,[3] in my view convincingly, that philosophers who followed Frege regarded the analysis of how language works, i.e., the creation of a theory of language (Dummett labels it "the theory of meaning"), as the central and primary task of philosophy. In fact the analytic branch of philosophy replaced epistemology with language analysis, making language analysis the kernel of all philosophy. As Dummett shows, this linguistic turn is connected to two features of Frege's views on sense. Firstly, it is a departure from a genetic approach; it purges logic from all questions related to the acquisition of senses of names, and asserts that these questions should be treated in psychology; (Dummett 1981a, 679):

> Frege diverted attention from the process of acquiring concepts[4] to the manifestation of possessing them. For human beings at least, the possession of a concept is manifested primarily in the employment of language. The analysis of concepts thus becomes an account of the working of language.

Secondly, even if there is a certain continuity with the preanalytic (epistemological) endeavor to analyze thoughts, analytic philosophy holds the view that thoughts (and senses in general) are given to us only through language, and in this way breaks from the preoccupations that marked preanalytic endeavors. This view can be found in various works of Frege. So in (1897) he claimed "[t]here is a difficulty here in that we think in some language or other" (1897, 142). Similarly he wrote on thought (1918a, 13):

> I am not here in the happy position of a mineralogist who shows his audience a rock-crystal: I cannot put a thought in the hands of my readers with the request that they should examine it from all sides. Something in itself not perceptible by sense, the thought, is presented to the reader – and I must be content with that – wrapped up in a perceptible linguistic form.

And finally he claimed (1924/25, 269):

> The connection of a thought with one particular sentence is not a necessary one; but that a thought of which we are conscious is connected in our mind with some sentence or other is for us men necessary. But that does not lie in the nature of the thought but in our nature. There is no contradiction in supposing there to exist beings that can grasp the same thought as we do without needing to clad it in a form that can be perceived by senses. But still, for us men there is this necessity.

So Dummett claims that what follows from Frege is that "philosophy of thought and of language are not merely related, but indissolubly entangled with one another: they can be pursued only simultaneously," (1981b, 50) that is to say "the only way in which we can give a philosophical account of thought is by giving an account of language" (1981b, 39). Dummett, by analyzing the works of Frege, frames analytic philosophy as follows (1981b, 39):

> The basic tenet of analytic philosophy, common to such disparate philosophers as Schlick, early and late Wittgenstein, Carnap, Ryle, Ayer, Quine, and Davidson, may be expressed as being that the philosophy of thought is to be equated with the philosophy of language: more exactly: (i) an account of language does not presuppose an account of thought, (ii) an account of language yields an account of thought, and (iii) there is no other adequate means by which an account of thought may be given.

This basic tenet of analytic philosophy had a profound implication for the very semantical investigation into the (Fregean) sense of language expressions. Because Frege in his *On Sense and Reference* did not provide a criterion of the identity for the logical entity he labeled "der Sinn," it was the task of the logicians, who came after him, to provide such a criterion. Because sense is viewed by analytic philosophy as an entity which can be approached only via the language expressing it, the logicians from this philosophical tradition attempted to provide a criterion of identity of sense of language expressions *not by investigating into the very sense*, but by means of the remaining two logical entities: the language expressions (designators) and/or the reference of these language expressions. So, in this philosophical tradition sense is viewed as a *non-authentic entity*, which can be approached by us only via language, as the *sole authentic logical entity*. Carnap, for example, in (1956) proposes a criterion of identity of sense of two language expressions by means of the identity of their reference in all possible worlds. In subchapter 10.3 I will show how, via the reconstruction of logical categories, it is possible to access the very sense of language expressions as an authentic entity.

Let me now pose the following question. Is there a place between psychology and a theory of meaning where epistemology can operate at all? Dummett, on the one hand, claims that the drive to replace epistemology by theories of meaning at the core of philosophy starts with Frege. With respect to this, Dummett mentions (1981b, 61) Wittgenstein's statement "Epistemology is the philosophy of psychology" (*Tractatus* 4.1121), and he views the very *Tractatus* as a "pure essay in the theory of meaning, from which every trace of epistemological or psychological consideration has been purged as thoroughly as the house is purged of leaven before Passover" (1981a, 679). But, on the other hand, Dummett complains that Wittgenstein's purge was "excessive" (1981a, 679). He mentions certain facts about Frege that prove that the latter was in fact opposed to such a purge. Frege claimed, e.g., that "knowledge as a mental process does not form the subject matter of theory of knowledge, and hence psychology is to be sharply separated from theory of knowledge" (Frege 1885, 102). Similarly, Frege wrote (1879/91, 3):

> Now the grounds which justify the recognition of a truth often reside in other truths which have already been recognized at all. But . . . this cannot be the only form that justification takes. There must be judgments whose justification rests on something else . . . And this is where epistemology comes in.

So, it seems to me, that Frege's answer to the question posed above would be positive. But then one faces another question, namely: What is such an epistemology about; what is its subject matter? It is neither the way man manifests his possession of senses (concepts) in the employment of language, nor his mental processes accompanying or preceding the acquisition of these senses (concepts). The former is a subject matter of the theory of meaning, while the latter a subject matter of psychology. An analysis of Frege's works shows that they contain neither an explicit nor an implicit answer to this second question.

Now my claim is that Frege could not have any answer to it. In fact, as I will now try to prove, the problem of justifying epistemology's right to exist has its roots at the very core of Frege's works. At this core, which informs his understanding of the subject matter of logic as science, lies his understanding of human beings. What he says about a human being, even if scattered throughout his various works, shows that his understanding is thoroughly psychologistic. He reduces human beings to entities endowed with consciousness, which he characterizes as the creations of sensations – imaginations, feelings, moods (1918a, 13–14) – which vary from individual to individual. So Frege reduces man to an isolated individual, to a lonely ego.[5] That this is so becomes apparent from the fact that when he mentions consciousness, he always writes about "my" or "his" consciousness (1918a, 14); in short, the owner, bearer of consciousness is an individual human being (1918a, 25). The reduction of human consciousness to the consciousness of an individual in turn leads to the reduction of human thinking to the thinking of an isolated individual, a process which goes on in the form of mental entities like ideas, representations, etc. The latter is exactly what Frege labeled "subjective performance of thinking" (1892, 62). But Frege also realizes that mankind possesses a common stock of knowledge which is passed from generation to generation, say "science common to many" (1918a, 17). Therefore, given his psychologistic understanding of man, human consciousness, and human thinking, Frege is forced *firstly*, to hypostatize that common stock of knowledge into an independent realm (*das dritte Reich*). *Secondly*, he is forced to view the realm of interindividual communication (the realm of language) as the only realm where we can scrutinize thoughts while escaping, if we are careful enough, the danger of being trapped in the web of the mental, psychic, subjective.

So, in my view, given Frege's psychologistic approach to man, man's consciousness and thinking, there is just no place for an epistemology positioned between psychology and theory of meaning. My analysis of Frege's understanding of man and its consequences shows also that the basic principles of analytic philosophy, as expounded by Dummett in the points (i) through (iii) given above, are only the necessary product of a psychologistic understanding of man.

For an epistemology such as science, independent from the theory of meaning and psychology, to be possible, it is necessary to start from a non-psychologistic understanding of human beings. My understanding can be outlined as follows. A human being is neither a behavioral entity processing sensory stimuli, nor is he or she a lonely ego whose thinking goes on in the form of private moods, feelings, images, etc., but is a cooperating, goal-setting, acting being, which through the cooperated action changes himself/herself and the (natural and social) reality. Therefore, I view the human subjective activity of thinking *not* as a process going on in the mind of an isolated individual who grasps outside him existing thoughts from the third realm, but as a subjective activity of many (cultures, generations, etc.) which creates, produces the very objective (i.e., also the intersubjective) content of thinking. I, therefore, also distinguish two types of science about the logical. Science about the logical – logistic and logical semantics – scrutinizes and explains the relations between the forms in which the interhuman symbolic communication proceeds and the laws which interhuman symbolic communication obeys (e.g., the relation of inference; how the sense of a language expression is coded by the context of a language expression into which the former is embedded). Subject to investigation here are the laws which safeguard the unambiguity of interhuman symbolic communication.

The science about the logical – *epistemology* – investigates the different relations and laws, namely those which are given between the forms in which human thinking goes on as an epistemic (cognitive) process. Subject to these is the very production of the objective content of human knowledge. These two types of relations or laws exist wherever and only wherever there are goal-setting beings which act in cooperation mediated by language.

I understand Frege's *Sinn* as cognitive sense, as the collectively owned result of the cooperative thought-activity of human beings, which, then, serves as the basis for a successive thought-activity, generating a new objective content of thinking, as well as for a successive practical-transforming activity. The possession of concepts (senses) by humans materializes, therefore, not only in their language (as Dummett claims), but also in their thought-activity, which produces new senses (thoughts), as well as in their cooperated action. The very process of human thinking has the character of a cooperating activity in which man creatively operates with an ideal object, with the subjective image of a real object. Thinking is understood by me as a system of thought-operations with an ideal object (object in mind or *thought-object*) by means of which it moves towards new knowledge. The cognition of a real object is viewed by me as existing when the mutual interaction between an object and a cooperative subject leads to the emergence of a naturally and socially conditioned, informative representative of a real object. So, cognition has the character of conscious reflection (sense, thought-object), where this conscious reflection is not merely a potential piece of information about a real object coded in another object,[6] but is an already interpreted piece of information, a conceptually thought object; the ideal being of an object in the subject.

When speaking about the forms of human thinking – as an epistemic, cognitive relation – which are the subject matter of epistemology, it is important to emphasize that these forms differ from the forms in which, say, thinking of physical sciences goes on. They are different from the sense of expressions like "wave function," "gene," etc., because their reference is not immediately a certain concrete object in reality (nature or society), but a general type of a thought-object, which is an invariant for any (natural or social) thought-object. In a culture where science is already present, they are the determinants and coordinators of scientific research, the framework which determines and determined the creation of scientific theories. They express what is common to special sciences of a certain historic period, what unites them, what is general and necessary for them without difference. Without those forms it is impossible to think a single object; man needs them for the creation of the ideal (thought-) objects of special sciences, expressed in empirical and theoretical terms.

My differentiation between two types of sciences whose subject matters are different types of logical relations (laws) does not mean that an intersection of these sciences is excluded beforehand. Because semantics scrutinizes the relation of language to the extra-language reality, where the latter contains an epistemic relation of the (cognitive) sense to the reference, more profound semantical considerations often face scholars dealing with issues which are of an epistemic nature. As an example of this let me take first Kripke's (1980) and then Carnap's (1939) paper. Kripke's semantical reflections on common names like "gold" are very closely related to issues which are labeled in this book as the "development and growth of knowledge" and which belong in the province of epistemology. Kripke first mentions an imagined case when mankind finds out that gold only appears yellow due to certain properties of the atmosphere in the regions were gold is mined and that gold is actually blue in color (1980, 118). Later on, people discover the microstructural properties of gold (its atomic number) and, on this basis, they are able then to explain the apparent properties of gold (color, metallic properties, etc.) as necessary with respect to that atomic number (1980, 125).

What Kripke in fact reconstructs here is the development of knowledge from phenomena (phenomenal properties) to the essence (essential properties), and then from the essence "back" to the phenomena. Of course this return is accomplished from an "over the top" theoretical location: seemingly, initially accidental properties are already explained and proved as necessary ones. Such a development and growth of knowledge can be traced also in Carnap's reflections in (1939) on the structure of scientific theories, already analyzed in this book in subchapter 1.4. When Carnap scrutinizes the relation of abstract terms ("electric field") to elementary terms ("yellow," "hard"), even if he prefers the "top-down" road of construction, i.e., the move which takes the abstract terms as the primitive ones, he still realizes that there exists also a move to the opposite direction (1939, 64):

> [H]istorically, science started with laws formulated in terms of a low level of abstractness. But for any law of this kind, one nearly always later found some exceptions and thus had to confine it to a narrower realm of validity. The higher the physicist went in the scale of terms, the better did they succeed in formulating laws applying to a wide range of phenomena.

So, for example, physics initially formulated certain laws for electric fields and other laws for magnetic fields. Physics then moved one step up and succeeded in the formulation of laws for electro-magnetic fields. From these laws it was possible to obtain, as special cases, laws initially designed to hold only for electric fields and laws initially designed to hold only for magnetic fields. Both groups of laws were proved to hold only under very special conditions.

The moral of Kripke's and Carnap's deliberations is that thinking as a heuristic, cognitive process, as a process of the production of (cognitive) senses in different special sciences, passes through similar forms, and that there should be a science which investigates these very forms. Analytic philosophy, which claims to be based on Frege's antipsychologism, but is in fact based on a psychologistic understanding of man, cannot provide a basis upon which such a science could be built. To stress this point, let me mention Frege's approach to what he labels "*Naturgesetze*." On the one hand, he views them as man's generalizations, as man's creations, with which natural occurrences are always in accordance (1918a, 1). But on the other hand, he claims (1897, 133):

> In order to be true, thoughts – e.g. laws of nature – not only do not need to be recognized by us as true: they do not have to have been thought by us at all. A law of nature is not invented by us, but is discovered.

These two claims hopelessly contradict each other. In order to eliminate this contradiction it is necessary to view thoughts as human creations, as results of human cognizing thinking and relabel Frege's "*Naturgesetze*" as "scientific laws." Epistemology, as already stressed, can be viewed as a discipline which reconstructs the forms of the production of senses, among them, of course, those which lead to the formulation of scientific laws, as well as those which follow from the formulation of scientific laws.

10.2 LOGICAL CATEGORIES

The fact that Kripke, when dealing with real human heuristic cognition, came across categories like "essence" and "phenomenon" is, at least in my view, only the consequence of the fact that the forms by which the human heuristic cognition, i.e., the process of production of thoughts, goes on are logical categories. Logical categories have been the subject of lengthy philosophical discussions continuing for thousand of years. According to Plato, they are ideas which are both kinds of concepts as well as the essences and archetypes of things (Plato 1961, 204). Aristotle claimed that categories are not the ideal essences of being, but are the highest genera of words (Aristotle 1941a, Chapter 4, 8) as well as the highest determinations of being (Aristotle 1941b, Book 5, 753–777).

According to Kant, categories are neither forms of thinking nor of being. They have the character of vacuous *a priori* logical forms which are themselves devoid of any content (pure); by means of these forms understanding (*Verstand*) realizes a conceptual synthesis of sensual representations (*Vorstellungen*), which have content but are formless, and which add to our judgments (in which the synthesis takes place) the character of necessity and universality. Accordingly, they are "keys to possible experiences" (A313/B370; Kant 1965, 310). Kant's ideas, unlike his categories, are forms of a higher theoretical synthesis of concepts. They are no longer forms of understanding but of reason (*Vernunft*).

Hegel returns to Plato's and Aristotle's view that categories are forms of being as well as of thinking, but he simultaneously develops a new approach to them as steps in the development of cognizing thinking. Categories are not external vacuous forms of thinking as it was claimed by Kant. They are "*determined*" because "they have a content" (Hegel 1840a, §43, Zusatz, 94; 1991, 86; 1959, 91).

According to him, in the case of logical categories, the notion "form of thinking" is equivalent to "abstract content of thinking." Logical categories, as the inner, non-vacuous forms of cognizing thinking, are not only the forms of understanding but also (and primarily) of reason. They are not only the principles of the conceptual synthesis of representations, but also of the theoretical synthesis of concepts. They have the character of logical determinations in the framework of which the concept develops from the subjective towards the objective, towards the theoretical and practical idea. Since they are non-vacuous forms of knowledge, in their own way they constitute knowledge, and they have their own objective subject matter. Naturally, only a general type of thought-object, representing and standing for an arbitrary thought-object of nature or society, corresponds directly to logical categories such as abstract knowledge (Hegel 1928, 101, 299; 1967a, 177, 436; 1969, 30-42; 1923, T. 2, 9–22). Because of this thought-object, i.e., *object* in thought, logical categories have an *objective content* simultaneously with a specifically subjective (human) form – object *in thought*. This is the *positive part* of Hegel's approach to logical categories. But there is also a *negative part* of it. Because he identifies being and thinking and reduces the objective reality existing outside thinking to the objective content of thinking, he breaks away that general type of thought-object (upon which he builds his *Science of Logic*) from reality (nature and society). Reality is understood by him as the objective content of a self-developing concept. As a consequence, he transforms cognizing thinking into the absolute idea, potentially involving the depicted reality; it becomes a substance-subject which is not only the ideal essence residing inside objects (as it was claimed by Plato), but also a demiurge of the natural and social reality (Hegel 1840a, §213, 385–388; 1991, 286–287; 1959, 352–353). He views the *Science of Logic* as "*the delineation of god as he is in his eternal essence before the creation of nature and a finite mind*" (Hegel 1923, T. 1, 31; 1969, 50). In his understanding, categories are steps in the logical genesis of the absolute idea. The ontogenesis and phylogenesis of the finite mind (intellect) is, according to Hegel, only an empirical manifestation of the logical genesis of the absolute concept, and only on this basis does it acquire the determination of freedom and credibility based on the cognized necessity (1840a, §12, 18–21; 1991, 35–37; 1959, 19–22). By appealing to the absolute idea Hegel attempted to substantiate the "axiomatic" character of logical categories, i.e., their ability to give the empirical cognition the seal of universality and necessity.

The central problem for any philosophical account to logical categories appears to be the problem of their origin. Hegel, in my view correctly, anticipated that they should be grounded in the "life" of the acting subject, in the contradictions between the subject and the object, in which the "external life" is gradually changed into the "inner life" (Hegel 1969, 769–772; 1923, T. 2, 423–426). But in Hegel's *Science of Logic* life (action) does not appear as sense-practical action of human beings, but only as thought-action, as a thought-plan of a future action. The origin of categories is substantiated by Hegel in the self-movement of pure reason, so that the latter formulates itself as the thesis, antithesis, and the synthesis. But real thinking cannot infinitely draw and derive its non-vacuous forms from itself. Because they are non-vacuous forms, and they have objective content, they can be drawn and derived in the long run only from the external world, from the objective reality existing outside cognizing thinking. The origin of logical categories should be searched for in the cooperated human activity of transforming objective reality. Logical categories are, relative to their origin, the generalized reflection of the schemes of practical action of the social man. The relations of things in the external world (their laws) are initially reflected in the form of thing-schemes of practical human action which are congruent with the schemes of things, of objects of this action (Ilyenkov 1977, 94–96).[7] After a certain time, the thing-schemes of practical action related to real objects are transformed into logical schemes of mind-activity related to the thought-object which are congruent with the schemes of practical action and things. Here the question of origin of the "axiomatic" character of logical categories finds its natural answer. Logical categories provide human thinking with a necessary and universal character because they are the quintessence of cooperated human practical action. They come into being through the generalization of the schemes of this action, but once given, they become the general way of seeing objects, the "instruction manual" for the creation of the schemes of theoretical and practical action.

The process of thinking, understood by me as a developing cognitive relation of the subject and object of cognition, goes on in thought-forms (*Gedankenformen*) labeled by me as logical categories. These should be understood as thought-expressions or abstractions of real determinations. Even though they are higher-level abstractions than, say the abstractions of the special sciences, they still constitute knowledge about objective reality, thought-expressions of real determinations. They are *not* thought-expressions of real determinations of nature by itself (these are given in the natural science), nor of the society by itself (these are given in the social sciences), nor of thinking by itself, but are the abstraction of the determinations of the real relations between thinking, nature and society, between the cognized being and the cognizing thinking. These relations are the essential conditions of their existence. Logical categories are given everywhere and only where the relation between being and cognition, subject and object exists.

What has previously been stated about forms of cognizing thinking holds for logical categories. The latter's object is not immediately a certain object in reality (nature or society), but a general type of thought-object which is an invariant for any natural or social thought-object. Logical categories appear with respect to concepts, say of the special sciences, as their abstract content. An abstract content with respect to a concrete content has the character of the latter's non-vacuous form. Logical categories can therefore be understood as non-vacuous forms of concrete thinking, as forms of the development of its content. Because this content is an objective one, its development represents the development of objective truth. I understand the latter as a unity of a content, which is independent from human beings (a *thing* in thought) and of a form which has a specifically human character (a thing in *thought*). Accordingly, since I understand logical categories as the thought-expressions of the real relation between cognizing thinking and cognized being (reality), and since this relation represents reflection of being into human consciousness, logical categories can be outlined as the general forms of the reflection and reconstruction of the objective in the subjective consciousness of man, as forms of his thought-grasping of objective reality.[8]

At this juncture two conclusions may be already drawn. First, logical categories are not eternal, because the determinations they reflect are not eternal; the latter are historical and temporary. In different periods of human history, society has solved different practico-cognitive problems by utilizing different logico-categorial "outfits," different networks of categories. This conclusion about the historicity of logical categories is important for the comprehension of the fact that they create historically becoming, developing, self-renewing (reproducing), and finally declining wholes.

Second, it is now possible to understand the threefold determination of logical-categories: 1) the *practico-ontological*$_1$ (the sense of this index will be evident below); the *epistemical-cognitive*; and 3) the *logical*.

Firstly, they are the reflection of the general forms of the existence of objective reality (nature and society), the generalization of the schemes of cooperated practical action (congruent with the schemes of objects); they provide human beings with an "instruction manual" and tool for interhuman interaction and for interaction of human beings with nature.

Secondly, they are the forms of the relation between the cognized reality and cognizing thinking. I distinguish here two principal forms of this relation. First, there are forms which represent the reflection of the objective in the subjective consciousness of human beings, i.e., forms of man's thought-appropriation of reality in accordance with reality's laws. There exist also forms which are different from the forms of conscious reflection, namely those which represent the forms of the *thought-transformation* of reality. Human beings through these forms create *projects* of their own future practical action. Logical categories are not only, as claimed above, general forms of reflection of the objective reality in the consciousness of human beings, forms of the grasping of the former by the latter, but also forms of the transformation of reality in thought in accordance with the aims, wants, and values of human beings. Marx labeled this form of the appropriation of reality as the *practico-spiritual* (*praktisch-geistige Aneignung*) form (1976, 37; Marx and Engels 1975ff., Vol. 28, 38). While in the process of reflection the aim of operations with the thought-object is to approach nearly as possible to the real object, in the practico-spiritual type of activity the aim is to create a type of a thought-object which, after its recreation in the real world, will satisfy the needs, wants, and desires of human beings. In the process of conscious reflection, the real object is given prior to our spiritual activity, but in the practico-spiritual activity it is the other way round. Marx expressed this as follows (1976, 37; Marx and Engels 1975ff., Vol. 28, 38–39):

The whole as it appears in the head as a thought-whole is a product of the thinking head which appropriates the world in a way which is the only possible one for it, in a way which differs from the artistic, religious, and practico-spiritual appropriation of the world. The real thing (Subject) still remains to exist in its independence outside the head so long as this head behaves only speculatively, theoretically. . . . [I]n . . . the theoretical method . . . the thing . . . has always to be before (*Vorschweben*) the representation.

Thirdly, they are the forms of the purposeful human activity with the thought-object, forms of the conceptual synthesis of representations and of the theoretical synthesis of concepts; the basis for the analytico-synthetic activity of thinking, moving from a less developed (cognitive) sense to a more developed one, the tool of the productive (heuristic) activity of thinking, of the creation of the objective content of thinking.

10.3 PHILOSOPHICAL CATEGORIES, SYSTEM OF PHILOSOPHICAL CATEGORIES, AND THE HISTORY OF HUMAN THINKING

Logical categories, I claim, come into being only where the relation of the subject and object is given. But only on a high developmental stage of human self-consciousness do the logical categories become the subject matter of a special analysis. They are investigated by philosophy, which theoretically expresses them as philosophical categories.

A correct understanding of the nature of logical categories, as well as the realization that they are one of the ultimate subject matters of philosophical investigations, becomes the central issue for philosophy after the emergence of special sciences. The last four centuries witnessed a dramatic formation of special sciences. Problems which were previously formulated within philosophy (e.g., What is life? What is the nature of the stars?) were separated from it and are now solved completely outside its framework by means of the special sciences. This in turn requires us to redefine the subject matter of the central philosophical disciplines – ontology, epistemology, and logic – as well as their mutual relations. If one understands *ontology* as a theory of the nature of being, then he faces the question: Did not the subject of ontology split up into the subjects of the special sciences? If one tries to understand logic as an integral part of philosophy, then the question arises: Is logic, after the emergence of symbolic logic as a special science, still a part of philosophy? In my view, the solution to these two questions should be found in a synthesis of a newly understood ontology, epistemology, and logic.

Firstly, the subject matter of ontology is not pure being, the universe as a whole, but only that part of it which is given in the practico-cognitive activity of humans. Therefore, the ontological question of the nature of being is transformed into the ontologico-epistemological question of the nature of the relation of being to consciousness.

Secondly, since the subject matter of epistemology is not only the content of thinking, but also the general forms of its movement towards new results, epistemology as science is also *logic* as science (I will label it "(philosophical) logic"), which is still different from symbolic logic. While the latter, together with semantics, investigates how the objective content of man's thinking is materialized in, and communicated by means of, linguistic forms, logic, which is a integral part of philosophy, investigates how the very objective content of thinking is produced. Exactly at this point the negative effect of Frege's psychologistic understanding of man becomes evident. On the one hand, he realizes that there exists an objective content of human thinking (*Gedanken*) which has a transindividual character. But because, on the other hand, a human being (the subject) is for him only a lonely ego, he views human thought-activity leading to the production of that objective content as a purely mental process, which, therefore, is not the subject matter of logic but of psychology. And, as a result of this – *in order to escape his own psychologism* – he projects that objective content of thinking into a realm (*das dritte Reich*) completely independent from man and his thinking. That this is so is readily seen from Frege's footnote in (1892), where he wrote "By a thought I understand not the subjective performance of thinking but its objective content, which is capable of being the common property of several thinkers" (1892, 62).

By reconstructing logical categories as philosophical categories it is possible to provide a criterion of the identity of sense of a universal name appearing in a scientific theory by targeting it as an authentic entity. In order to make my point more understandable, let me first return to Marx's thought movement from exchange-value via value "back" to exchange-value as reconstructed by me in Chapter 6 above. What one is dealing with here is in fact a change of the (Fregean) sense of the name "exchange-value," i.e., this name has in the case of Marx's thought movement in the first volume of *Capital* two different senses. As shown already in Chapter 6, this movement from the former to the latter can be characterized as a movement of knowledge from appearances to manifestations. These two categories, as shown already in chapter six, also characterize Newton's movement from Definition II through VII to the three laws (*Lex I*, *Lex II*, *Lex III*) in his *Principia*. What is especially noteworthy here is the change of the sense of the expression "motion." Newton introduces it for the first time in Definition II as a magnitude given by the multiplication of mass and velocity. Then, in the three laws this magnitude is characterized as being determined by the action of forces. Its change or immutability is explained via forces (their action or non-action). Initially this magnitude has the character of a kinematic magnitude; finally that of a dynamic magnitude.

The fact that I characterize the change of the (Fregean) sense of universal names like "exchange-value" or "motion" by means of logical categories of appearance and manifestation, enables me to provide a criterion of the identity of sense of a universal name appearing in a scientific theory, while simultaneously *bypassing the name's reference*, i.e., to provide a criterion of identity of sense by approaching the latter as an *authentic entity*. In order to determine whether the sense of a universal name from a scientific theory has changed or not, one has to investigate into the categorial "outfit" of that part of this scientific theory, where that sense is constituted. If the categorial localization of the sense of that name has not changed inside this theory, or among two (or more) theories, then this sense has not changed as well. If, as it is, for example, in the case of the sense of the name "exchange-value" in the first volume of Marx's *Capital*, or in the case of the sense of the name "motion" in Newton's *Principia*, that categorial localization has changed, then the sense has changed as well.

From what was stated above it follows that philosophical categories should be formulated in the unity of ontology$_2$, epistemology, and (philosophical) logic. Only such a synthesis can theoretically grasp the threefold determination of logical categories (practico-ontological$_1$, epistemical-cognitive, logical), and safeguard philosophy's survival after the emergence of the special sciences. The investigation of logical categories is, in my view, a specific task of philosophy which cannot be substituted by any other discipline. Its ultimate task appears to me to be the reconstruction of the whole network of logical categories. In the European philosophical tradition, the most advanced attempt at such a reconstruction was made by Hegel. What is worth mentioning with respect to Hegel's attempt is the method he uses in this reconstruction. A reconstruction of logical categories has to grasp them as changing and developing, in their origin and genetic connections, and reconstruct both the dynamics of individual categories as well as the wholeness of their mutual relations. In my view this can be fulfilled by a system of philosophical categories built by the method of proceeding from the abstract to the concrete. This method can reconstruct the network of logical categories as a system of thought-forms going on through a continuous process of development (i.e., as a system continuously becoming anew and reproducing itself). Such a system has a synchronous-diachronous character. Logical categories are grasped by this method in their origin and changing, in their genetic connections with other categories. Such a system makes clear not only the relation of coordination but also of subordination. It shows how through these relations the content of one category is passed over to the whole network, so that each category is connected to all the remaining. Each less developed category is portrayed by this method as the *presupposition* of a more developed one and simultaneously as its *result*, because it is reinterpreted and transformed on a deeper basis. In a system of philosophical categories built by this method, each step forward is therefore an unfolding of the previous one and simultaneously its substantiation; each step forward is seemingly a return to the starting point. In a system built by this method, individual philosophical categories, as well as their unification into clusters, theoretically reconstruct the steps in the development of cognizing thinking.

The method of proceeding from the abstract to the concrete was delineated by Hegel (1969, 825–844; 1923, T. 2, 483–506) and applied in the construction of his system. It starts from a category which, according to Hegel, is the most simple, immediate and abstract ("being"-"nothing") and proceeds to the concrete totality of the theoretical and practical idea. In this process each succeeding category (polarity of categories) preserves and overcomes the previous. Similarly, his system develops in such a manner that after a certain number of steps it returns to the initial point of departure and gains by this process, for it has been enriched by the mediating determinations (Hegel's triad). Accordingly, his whole categorial system resembles a big circle composed of smaller circles. The big circle is expressed by categories "being," "essence," and "notion." Each point on its circumference is in its turn also a big circle and so the big "coil" unwinds itself as follows: quality, quantity, measure;

essence, manifestation, reality; subject, object, idea. By means of the method of proceeding from the abstract to the concrete, Hegel processed over two hundred categories. With respect to this positive moment as well as to those mentioned earlier, any attempt at a construction of a system of philosophical categories should be judged according to Hegel's *Science of Logic*.

But one has to bear in mind that the *Science of Logic* as a whole cannot be utilized as a basis for thought-operations, say, in a special science. Hegel purportedly built it not as a theoretical reconstruction of the logical forms of real human thinking, but of the self-determinations of the absolute idea. The development of philosophical categories in the *Science of Logic* expresses, according to Hegel, the self-movement of the self-forming idea which is born from nothing, and which outside itself "*may not presuppose anything*" (Hegel 1923, T. 1, 54; 1969, 70).[9] The *Science of Logic*, therefore, represents a system of philosophical categories of idealistic dialectics. In chapter eleven I will analyze how this idealistic framing penetrated into the very categorial clusters of his *Science of Logic*. In my view the system of philosophical categories should from the very beginning be built as a reconstruction of the development of human cognitive forms of practically acting cooperated human beings, as a reconstruction of the development of the forms of the spiritual and practical appropriation, "seizing" of the (natural and social) reality, i.e., as a system of categories of realistic (materialistic) dialectics.

All the features of Hegel's idealism mentioned till now enable me to compare Hegel's approach to the objective content of thinking with that of Frege. So as Hegel breaks cognizing thinking away from human beings and transforms it into the absolute idea, which only through (self-) genesis becomes the finite (human) intellect, he is capable of scrutinizing – even if only in such a mystified way – the objective content of human thinking independently from its materialization in language. Therefore, even if Hegel did not intend it, his approach to the objective content of thinking (thoughts) should be evaluated at least on a par with Frege's approach, which analyzes thoughts by means of an analysis of their materialization in language forms.

My earlier characterization of a system of philosophical categories can be further clarified. I claimed that 1) logical categories in the long run have their origin in the praxis of man, and that 2) logical categories set up a developing system with non-linear relations among them. These two determinations place certain constraints on the construction of a system of philosophical categories. Firstly, a system of philosophical categories should express the movement of cognition as starting from practical action and finding its (relative) end again in practical action. Secondly, it should express the mutual relations of the developmental stages (phases) of cognition in a non-linear manner, i.e., that the latter stages transform the former into their own result and substantiate them anew. These requirements lead to the scheme:

$$Pr \rightarrow E \rightarrow T \rightarrow PS \rightarrow Pr'$$

which I, following Černík's (1977; 1986), name as the scheme of the reproduction cycle of cognition. Pr denotes praxis, E the empirical stage of cognition, T the theoretical stage of cognition, PS the stage of the practico-spiritual activity, Pr' a more developed – with respect to Pr – praxis, because it is already mediated by E, T and PS. $E \to T$ expresses the descriptive-explanatory phase of cognition; man here in mind approaches the object. The PS-stage represents the prescriptive phase; man here in mind lets the object approach him. In mind he shapes the object in accordance with his values and relations of preference. In this phase the knowledge acquired in the descriptive-explanatory phase is subordinated, by means of a plan of a future transformation of the object, under man's interests and aims.[10]

The concept of the reproduction cycle expresses the mutual relations of these forms of human activity in a non-linear way. So, for example, empirical cognition which is initially the presupposition of theoretical cognition is then reinterpreted on the basis of this cognition. So the scheme can be written as

$$Pr \to E \to T_{E'} \to PS_{E''T'} \to Pr'$$

where E' denotes empirical cognition reinterpreted on the basis of theoretical cognition; E'' and T' denote the forms of cognition reinterpreted on the basis of the practico-spiritual activity. This scheme views cognition as deprived of a "pure empirical" or a "pure theoretical" basis, as a multidirectional movement whose components are moments in a non-linear developmental cycle, periodically opening and closing itself to praxis. My *first* additional requirement for the construction of a system of philosophical categories is that it should cover all the stages of the reproduction cycle of cognition.

I also claimed that the system of philosophical categories should express the sequence of operations with the thought-object, and I claimed that logical categories represent what is common, necessary, and unifying to all (thought-objects) of the special sciences in a certain time. From this it might appear that I am claiming that science as a whole in a certain historic period is completely uniform, i.e., that all disciplines of special sciences as well as different theories in one discipline in the same time have an identical logico-categorial apparatus. But the truth is that they use thought-objects of different degrees of complexity, that they have logico-categorial "outfits" of a different richness. This puts an additional constraint on the construction of a system of philosophical categories. Suppose one would construct a system which at most could serve as the framework for thought-operations with particles in classical mechanics. Such a system would show up as insufficient, as too poor to serve as the basis for operations with a thought-object displaying non-linear behavior, say, the hypercycle in molecular biology or the Brusselator in physical chemistry.

My second additional requirement for the reconstruction of a system of philosophical categories can be stated as follows. Any attempt to reconstruct the logical categories of a certain historic period (epoch) presupposes finding a certain type of a thought-object which, in this period, is the most developed, to which corresponds the richest logico-categorial apparatus. From a system of philosophical categories which would grasp this richest "outfit," it would then be possible to "peel out" those individual categories and clusters of categories which suffice for the investigation of simpler, less developed thought-objects of this epoch.

I understand the network of logical categories as a historically produced and declining system of the thought-appropriation and thought-transformation of the world, and I always refer to logical categories of a certain historic period. Such a system, in my view, contains in itself some of the categories of the preceding historical periods reinterpreted and transformed on the system's own basis. The philosophical reconstruction of a certain network of logical categories can, therefore, uncover some of the determinations of the historically preceding types of thought-appropriation and thought-transformation of (natural and social) reality; it can serve as an "anatomic key" to them.

This fact was realized by Hegel but simultaneously misrepresented by him. In his view "logical philosophy" (his system of categories) expresses the self-movement of the pure idea outside time and space. Historical philosophy investigates this movement in time and space. It reconstructs the empirical form of the logical genesis of the idea in a certain place, in a certain nation. The historical is, according to Hegel, the logical in its empirical form, and, therefore, the logical progression of categories is by-and-large identical with their historical progression, and the progression of philosophical systems in history is "similar to the sequence in the logical deduction of the Notions-determinations in the Idea" (Hegel 1840b, 43; 1955, Vol. I, 30). He claims that (1840a, §13, 21; 1991, 37–38; 1959, 22–23)

> [i]n the proper form (*eigenthümliche Gestalt*) of *external history*, the origin and development of philosophy is represented as the *history of this science*. This form gives to the developmental stages of the idea the form of an *accidental* succession and seemingly of a mere difference of the principles and their realization (*Ausführung*) in their philosophies. But the architect (*Werkmeister*) of this work of millennias is the one living spirit. Its thinking nature is to bring to its own consciousness that *what he is*. By becoming in such a way an object for itself, it simultaneously raises itself above it and reaches a higher stage of itself. *The history of philosophy* presents in the various appearing philosophies, partially, only *one philosophy* at its various stages of formation (*Ausbildung*), and, partially, that the particular principles, which are the basis of each philosophy, are just the *branches* of one and the same whole.

But the real relation of the logic of thinking to its preceding history is different. In the development of human thinking qualitative changes, transformations, breaks take place, which are characterized by a complete rebuilding of the whole network of logical categories. These breaks represent a global change of the whole style, the habitus of thinking, a change of rationality. Society, originally operating within a certain framework of thinking and practical activity, finds itself after a certain time at a stage where this framework is insufficient. The dominating (paradigmatic) network of logical categories cannot be used any more for the newly arisen cognitive and practical problems faced by the society. The existing network of logical categories has to be enriched by introducing new categories or a cluster of categories, with some of the original categories and clusters of categories partially losing their importance (a "dwarfing" of categories) or even losing it completely, and so dropping out from the apparatus of the rationality of that historic period. I claim, contrary to Hegel, that a seemingly identical category in different logico-categorial networks can in fact have different senses, and therefore also fulfill different functions in the history of human thinking. Accordingly, the very history of human thinking should be understood as a process of successive qualitative changes of categorial networks, each representing a different way of "seeing" and transforming the (natural and social) world, a different style of thinking and practical activity.

Everything said till now about logical and philosophical categories, about the reproduction cycle of cognition, and about the system of logical categories enables me on the one hand to view the reconstruction of scientific laws and of scientific explanation in a broader epistemological setting, namely in the context of the investigations of categories. On the other hand, my typology of scientific laws enables me to make more precise the very idea of the reproduction cycle of cognition. Even if I spoke above about the reproduction cycle of cognition, my differentiation between the pure idealizational type of a scientific law and the inherent type of a scientific law as two types of a scientific *theoretical* law[11] necessitates a distinction between two types of the reproduction cycle of cognition. While the one is "centered" around the pure idealizational type of a scientific law, the other is "centered" around the inherent type of a scientific law. These two types of the reproduction cycle differ from each other firstly by the categorial "outfit" of their theoretical stage (T-phase). But this in turn means that these two types of the reproduction cycle differ from each other also by the categorial cluster of their practico-spiritual stage (PS-phase). The PS-phases (and therefore also the Pr'-phases based on them) have to differ because they represent a thought-transformation of the object. A thought-transformation based on the pure idealizational type of a scientific law can plan a man's practical action which, at most, will be congruent with the determinations of objects of the real world. As an example of this let us take the law of gravitation. As shown in subchapter 6.2, on its basis we can calculate the size of the escape velocity which a body has to reach in order to escape the gravitational field of a planet. Because in theoretical mechanics we do not know how gravity as force is being produced (what are the essential conditions of its (re)production) in our practice of space flights we have to accept that we cannot abolish gravity in a certain location, and can, therefore, escape it only by accelerating a body (e.g. a space probe) to its escape velocity. But a thought-transformation based on an inherent type of a scientific law can push further and plan a praxis in which man will create and abolish objects in their very essence, a praxis where man practically will generate objects from their essential conditions. For example in the case of gravity the knowledge of the conditions of its very (re)production would give us, at least potentially, the recipe how in practice we could abolish gravity in the location where our space probe is located.

I leave the issue of whether those two reproduction cycles differ also in their respective categorial clusters of the empirical stage (E-phase) unresolved. On the one hand, it seems that the categorial clusters of the E-phase have to differ because this is only the preparatory phase for two different types of a T-phase. Categories of empirical knowledge which are the basis for the formulation of the pure idealizational type of a scientific law should be different from categories of empirical knowledge leading to the formulation of the inherent types of a scientific law. But, on the other hand, my analysis from chapter 6 showed that knowledge fixed in the pure idealizational type of a scientific law (e.g., appearance of the immanent measure) is the starting point for the movement of knowledge towards the inherent type of scientific law (grasping of the immanent measure and the explanation of the genesis of the manifestation of the immanent measure). So I cannot exclude *a priori* the option that there is no specific categorial cluster of empirical knowledge corresponding exclusively to the *T*-phase of scientific knowledge where inherent types of scientific law are formulated.

CHAPTER 11

HEGEL'S AND MARX'S CATEGORIES: A COMPARISON

The aim of this chapter is, by comparing Hegel's approach to logical categories with those used by Marx in his economical works, to show to what degree Marx relied on Hegel's *Science of Logic* and *Encyclopaedia Logic*, as well as to show the points of their principal disagreements.

I start by showing the importance of Hegel's philosophy for contemporary philosophy of science. I then relate this positive aspect of Hegel's works to his critique of Kant's *Critique of Pure Reason*. Finally, I show the differences between Hegel and Marx, by analyzing the idealism of the former and the realism (materialism) of the latter.

11.1 HEGEL AND CONTEMPORARY PHILOSOPHY OF SCIENCE

What makes Hegel at first glance so relevant with respect to contemporary philosophy of science and epistemology is his approach to the concept of scientific law as given in the *Phenomenology of Mind*.[1] There he explicitly reconstructs various types of scientific law, which in turn represent successive developmental stages of human knowledge. What makes Hegel's *Phenomenology* also highly interesting is the fact that it contains an implicit reconstruction of the cognitive structure of scientific laws.

Hegel first introduces the concept of scientific law at the level of the understanding consciousness (*verstehendes Bewußtsein*). The latter emerges from the sensuous and formulates there the first type of a universal (the absolute universal), which is the unity of a manifold of sensuous singulars. This unity has, according to Hegel, the character of a *force*. The goal which the understanding aims at is "to bring out the 'unconditional universal' . . . as the *inner* (unperceived) being of things" (1928, 106; 1967, 185). Understanding uses the ingenious trick of reflecting upon the mutually opposite action – interplay – of two forces, to look into the background of things (1928, 110; 1967, 190). In such a way not only the outside, the external features of things, but also their inside (*Inneres*), becomes the object of understanding and a "supersensible world . . . opens up as a true world lying beyond the sensuous world . . . over the vanishing *hereness* (*Diesseits*) the permanent *beyond* (*Jenseits*)" (1928, 111; 1967, 191). According to Hegel, the most elementary form of that interplay of forces has the character of the *law of force* (1928, 114; 1967, 195). This law, because it unifies the various unstable, perpetually changing, sensuous phenomena into a stable unity, turns the supersensible world into a "quiescent (*ruhiges*) realm of laws . . . beyond the world of perception" (1928, 114; 1967, 195).[2]

In the *Phenomenology* Hegel approaches the concept of scientific law for the second time at the level of self-consciousness and relates it to the *reason which observes nature*. Here again, as at the level of consciousness, he treats it in the context of relation of the sensuous hereness to the universal as something identical-with-itself (*Sichgleichbleibendes*) (1928, 185; 1967, 284). What comes in here as a new moment, and what distinguishes this level of knowledge from the knowledge at the level of understanding consciousness, is differentiation between the *essential* and the *non-essential* properties of things (1928, 187; 1967, 286). The former are the *quiescent* or *peaceful* determinations (*ruhende* Bestimmtheiten) (1928, 189; 1967, 288), and the instinct of reason tries to discover them.

The immediate importance of Hegel's approach to the concept of scientific law at the level of consciousness and reason is readily seen. Hegel views scientific laws as a type of knowledge which tries to go beyond the level of the immediately (sensuously) observable, the singular, as a way to assign to knowledge the determination of universality, this latter characteristic being attainable exclusively beyond the realm of the observable.

What is also worth noticing is that Hegel investigates the operations by means of which knowledge gradually rises above the level of the sensuous. Hegel labels these operations "purification of the law." He claims that reason (1928, 191; 1967, 291)

> experiments with the law. As the law first appears, it represents itself impure, enveloped in the singulars of the sensuous being . . . The instinct of reason aims in its experiments to find out what happens in such and such circumstances . . . this investigation has the inner purport to discover the *pure conditions* of the law.

What changes, according to Hegel, as a result of this type of thought-operation is, firstly, the *universe* for which the law is formulated. He claims, for example, that positive and negative electricity were initially related to resin and glass, and later, due to those operations, one obtained "pure" negative and positive electricity (1928, 191, 191; 1967, 291). Secondly, what can be gained by means of these operations is a type of scientific law different from the type of law subject to the operations. Hegel labels this result the *pure law* (1928, 192; 1967, 293).

The process of purification of a scientific law can be made more understandable by dealing with the development of Ohm's law as reconstructed by Izabella Nowak in (1974). Initially it had the following form:

$$L : (x)(Conx \rightarrow Ex = RxIx) \tag{11.1}$$

where "*Con*" denotes the set of conductors; "*E*" denotes voltage; "*R*" denotes resistance and "*I*" denotes electric current. But the actually measured values of these magnitudes did not correspond to those predicted on the basis of the equation $E = RI$ from (11.1). This equation, it was later discovered, would hold only if the capacitance, *C*, of the capacitor given in the circuit would be infinite. As result of this discovery the following form of Ohm's law was obtained:

$$L^{(1)} : (x)(Conx \,\&\, 1/Cx = 0 \rightarrow Ex = RxIx) \tag{11.2}$$

If "*Q*" denotes the charge of the capacitor, then $L^{(1)}$ can be concretized as follows:

$$L^{(0)} : (x)(Conx \,\&\, 1/Cx \neq 0 \rightarrow Ex = RxIx + Qx/Cx) \tag{11.3}$$

But again the results of measurements somehow differed from the theoretical calculations based on (11.3). Therefore, neither $L^{(0)}$ can be used for the explanation of data obtained by means of measurement. Later it was discovered that the failure of $L^{(0)}$ has its origin in $L^{(1)}$. The equation $E = RI$ holds only if yet another idealization is included: the electromotive force, e, is equal to zero. One then obtains as follows:

$$L^{(2)} : (x)(Conx \,\&\, 1/Cx = 0 \,\&\, ex = 0 \rightarrow Ex = RxIx) \tag{11.4}$$

The gradual concretization based on it yields:

$$L^{(1)} : (x)(Conx \,\&\, 1/Cx = 0 \,\&\, ex \neq 0 \rightarrow Ex = RxIx + ex) \tag{11.5}$$

$$L^{(0)} : (x)(Conx \,\&\, 1/Cx \neq 0 \,\&\, ex \neq 0 \rightarrow Ex = RxIx + ex + Qx/Cx) \tag{11.6}$$

But once again the results of measurement diverged from the results of calculations based on (11.6). It was discovered that $E = RI$ holds only if one includes yet another idealization: self-inductance, L, is equal to zero. One then obtains the following scientific law:

$$L^{(3)} : (x)(Conx \,\&\, 1/Cx = 0 \,\&\, ex = 0 \,\&\, Lx = 0 \rightarrow Ex = RxIx) \tag{11.7}$$

Gradual concretization based on (11.7) leads to the following scientific law

$$L^{(2)} : (x)(Conx \,\&\, 1/Cx = 0 \,\&\, ex = 0 \,\&\, Lx \neq 0 \rightarrow Ex = RxIx + LxdIx/dtx) \tag{11.8}$$

$$L^{(1)} : (x)(Conx \,\&\, 1/Cx = 0 \,\&\, ex \neq 0 \,\&\, Lx \neq 0 \rightarrow Ex = RxIx + LxdIx/dtx + ex) \tag{11.9}$$

$$L^{(0)} : (x)(Conx \,\&\, 1/Cx \neq 0 \,\&\, ex \neq 0 \,\&\, Lx \neq 0 \rightarrow Ex = RxIx + LxdIx/dtx + ex + Qx/Cx) \tag{11.10}$$

The whole network of relations between the developmental stages (11.1) through (11.10) of Ohm's law can be represented by the following scheme:

$$
\begin{array}{cccc}
L_1 \Rightarrow & L_3^{(1)} \Rightarrow & L_6^{(2)} \Rightarrow & L^{(3)} \\
& \perp & \perp & \perp \\
& L_2^{(0)} \Rightarrow & L_5^{(1)} \Rightarrow & L^{(2)} \\
& & \perp & \perp \\
& & L_4^{(0)} \Rightarrow & L^{(1)} \\
& & & \perp \\
& & & L^{(0)}
\end{array}
$$

Here "$\Rightarrow$" stands for the discovery of a previously unknown (hidden) idealization; "$\perp$" stands for the relation of gradual concretization and the lower numerical indexes indicate the order of the realization of a failure.

What has to be emphasized here is that on the one hand, because the laws of the lowest degree of idealization (here $L_2^{(0)}$, $L_4^{(0)}$ and $L^{(0)}$) serve as the basis of explanation of what is going on in experiments and measurements, the realization that the concretized laws, and sometimes also the laws of the highest degree of idealization (here $L_3^{(1)}$, $L_6^{(2)}$ and $L^{(3)}$) have failed, enters into the above reconstructed explanation-cum-concretization hierarchy from its "bottom." But on the other hand, the process of explanation by gradual concretization starts from the "top" of this hierarchy. For example, according to the second column of the scheme, the experimenting and measuring scientist realizes that $L_2^{(0)}$ – which, before this realization, was obtained by means of a "top-down" explanatory procedure – has failed in respect to certain measurements and then, after a close scrutiny of it, proceeds, if necessary, "up" and investigates into $L_3^{(1)}$. These two opposite directions of a thought movement are inseparably entangled. In order for the laws at the "bottom" of the concretization hierarchy causally to explain what is going on in measurement and experimentation, they have to be derived from the "higher"-level causal laws of this hierarchy. But on the other hand, in order to discover and state all the laws in this hierarchy – and this includes the discovery of previously unknown (hidden) idealizations – we need also the movement of cognition from the "bottom" to the "top" of the hierarchy. In subchapter 3.4 I claimed that the process of explanation by gradual concretization, i.e., the thought movement from the "top" to the "bottom," has the character of a heuristic process creating new knowledge of the causal genesis of phenomena. The same can be stated about the thought movement going in the opposite direction. The information about the results of measurement and experimentation enter the hierarchy at its "bottom." From there it moves gradually "up," level by level, where it is processed – by means of a discovery of hidden idealizations – in order to provide a causal understanding of the results (near failure or complete failure) of the causal explanations.

What also has to be emphasized here is that the above given reconstruction of the process of purification does not cover one important aspect of Hegel's ideas about it, namely, that the very universe of discourse can change in that process. In I. Nowak's reconstruction the whole growth of knowledge is bound to one kind of entity: conductors. The case of a growth of knowledge where the very universe of discourse changes is not reconstructed by her.[3]

How does Hegel's approach to the process of purification differ from that of J. L. Mackie in (1980) which we dealt with in subchapter 2.6? The latter, in my view, uses the idea of purification primarily as an instrument of a philosophical (metascientific) investigation into scientific laws already formulated by special sciences. Hegel, on the contrary, views this type of thought-operation as a means for the creation of new knowledge fixed in scientific laws. By means of them cognition transforms certain scientific laws into different ones with a deeper epistemic (cognitive) depth. The latter, initially the result of cognition, in their turn become the object (*Gegenstand*) (1928, 192; 1967, 293) of yet another, higher-level, more developed type of "observation." Exactly in this sense Hegel claims that the pure type of scientific law develops into the organic type of scientific law.[4]

What makes Hegel's approach to the concept of organic law worth mentioning is the fact that it is from the point of view of the *sides of the law* (*Seiten des Gesetzes*) (1928, 193; 1967, 294). From this point of view he investigates the various possible types of sides which could be related in an organic law: organic nature and the inorganic nature; the internal and the external of the organic; and the relation of the aspects of the internal of the organic (sensibility, irritability, reproduction). By analyzing all possible options of the relations between the sides of an organic law, he finally arrives at the conclusion that "the representation of a law is lost altogether in the case of the organic" (1928, 207; 1967, 310).

Despite this claim, the importance of Hegel's concept of an organic type of law is that he approaches it via the relation of the sides of a law. He explicitly speaks about aspects related in a scientific law which do not have the character of empirically observable events, but have a *highly abstract, unobservable, theoretical* character.

In sum, one can see that Hegel's treatment of the pure and the organic type of scientific law results, *firstly*, in the reconstruction of its most important constituents: the *universe* for which it is formulated; the *conditions* under which it should hold; and finally a *certain type of relation* of the aspects, the sides fixed by the law. *Secondly*, Hegel treats the category of scientific law as something characterized by stages. It appears at a certain level of knowledge, and itself passes through various stages. These stages evolve one from another due to the *productive character of thinking*, so that the next stage is, as compared to the previous one, always a deeper, more advanced form of knowledge.

The fact that Hegel was able to address these two issues, which still have actual bearing for contemporary philosophy of science, can be understood primarily *vis-à-vis* his discussion and critique of Kant's *Critique of Pure Reason*.

11.2 HEGEL vs. KANT ON UNDERSTANDING AND REASON

The *Critique of Pure Reason* (*CPR*), understood by Kant as a critical investigation into the powers of intellect as a source of putative knowledge, introduces several new approaches into philosophy. For one thing, Kant investigates the very character of metaphysical categories which were in pre-Kantian philosophy used as 'tools' for grasping reality without ever subjecting them to investigation. Kant, by treating these categories as specifically human forms of encountering reality, triggered the process of the creation of a new philosophical discipline, namely, that discipline characterized by the unification of logic, metaphysics and, as we will see below, epistemology.

This new moment is related to another aspect of Kant's *CPR*, namely its new approach to the understanding of what logic is (or should be).[5] Kant divides the science of logic (*Wissenschaft der Logik*) into a *general logic* (*Logik des allgemeinen Verstandsgebrauchs*) and into a *special logic* (*Logik des besonderen Verstandsgebrauchs*) (A52/B76; 1965, 93). The first can perform its work without any special knowledge of the objects. It divides itself further into the *pure general* (*formal*) *logic*, and into the *applied general logic*. The latter takes into account the accidental conditions under which rules of formal logic are used by humans, like force of habit, inclination, prejudice, etc., i.e., it has the character of a *psychology of thinking* (A54–55/B78–79; Kant 1965, 94). The pure general logic investigates the "absolutely necessary rules of thinking," (A52/B76; 1965, 93) which it has to follow in order to be correct. It serves exclusively as a "canon of judgment" (A61/B85; 1965, 99). It does not investigate the thinking which has the status of *cognition* (*Erkennen*), i.e., when there exists, or can exist, an object related to thinking. Thinking which is not cognition is required only to be free of contradictions (Bxxvii; 1965, 28), and this type of logic "has the modest merit of guarding against error" (A795/B823; 1965, 629).

The special type of logic, contrary to the pure general type, is in fact a set of logics, because it depends on the ways that the proofs and arguments are performed in individual special sciences. What all these special logics still have in common is the fact that each of them fulfills, with respect to is own science, the task of an *organon* (instrument of cognition). Their purpose is: "to extend and enlarge our knowledge" (A61/B86; 1965, 99). What the *CPR* in fact then pursues is an investigation into the suppositions for the creation of *transcendental logic*, which is the logic of a special science, namely, metaphysics.

Another new moment of Kant's *CPR* is given by the fact that he departs from the preceding philosophical tradition, which interpreted differences between representations, concepts, and ideas as mere differences in their coherence, clarity, or adequacy.[6] Kant alters this understanding of human cognitive capabilities in a way that leads him to introduce a fundamental type-differentiation between representations and concepts as well as between concepts and ideas. He distinguishes between the *representations* (*Vorstellungen*), the *concepts of understanding* (*Verstandsbegriffe*), and the concepts of reason (*Vernunftsbegriffe*); the latter being ideas proper.

The importance of Kant's distinction between representations, concepts of understanding, and concepts of reason is, *firstly*, that he turns the concepts of understanding – *categories* – into the subject matter of a special investigation. He correctly, in my view at least, claims that without categories we cannot think a single object of our sensuous experience (A125, B143; Kant 1965, 147, 160), i.e., "[w]e cannot think an object save through categories" (B165; 1965, 173). According to Kant, the investigation into categories is central to understanding why human knowledge has the status of universality and necessity. My earlier claim, that Kant in his first *Critique* redefines the notion of logic as science, becomes more clearly understandable here, as we note that Kant broadened the scope of logic so that the very *categorial schemes of human cognition*, by means of which our stock of knowledge grows, become the subject matter of its investigation.

The importance of Kant's distinction between representations, categories, and ideas is, *secondly*, given by the fact that he subjects to investigation the *higher-level synthesis of the concepts at the level of reason* and regulative role of ideas – as concepts of reason – in this type of process.[7]

These two positive aspects of Kant's *CPR* distinctively contrast with its negative and, in fact, paradoxical aspects. Firstly, even if, according to Kant, categories assign to knowledge the status of universality and necessity, i.e., they constitute a *new content* absent at the level of sensuous representations, still he claims that categories are completely subjective and devoid of any content.

Secondly, he argues that because categories are empty and acquire their content only via sensuous representations, they are inherently restricted to the empirical realm. We are not allowed to use them in a transcendental sense, i.e., beyond the sphere of the empirically given (A219; 1965, 239). Beyond this sphere categories lack the element of sensuous intuition (*sinnliche Anschaung*). Therefore, all attempts to form a synthesis by their means, independent from experience, are bound to fail.

Thirdly, according to Kant, there is an unbridgeable gap between the *phenomenon* and the *noumenon*. Even if categories enable us to synthesize phenomena given to us at the level of sensuous representations, where these phenomena are somehow related to the object of cognition, and so enable us to *think this object* at all, still we should not be able to recognize what the object is beyond the sphere of the phenomena. Expressed otherwise: "though we cannot *know* . . . objects as things in themselves, we must be yet in the position to *think* them as things in themselves" (Bxxvi; 1965, 27).

Fourthly, the knowledge at the level of reason has a purely negative (illusionary) character. Reason discovers that it is incapable of grasping the unconditioned (god, soul, world). All attempts to grasp the unconditioned conclude with reason's entanglement in a net of antinomies. Ideas of reason do not enlarge the stock of our knowledge. They do not fulfill the function of a constitutive principle of reason (like categories for understanding), but are only the regulative principles of the systematic unity of the manifold of empirical cognition (B699; 1965, 550).

Fifthly, even if ideas are concepts of reason, Kant, in order to determine them, ultimately draws upon the objects – god, soul, world – of the same metaphysics which he himself criticizes.

These negative aspects of Kant's approach to pure (speculative, theoretical) reason, result in a *strict separation of logic from metaphysics at the level of understanding* (at the level where one can acquire at least the *knowledge* of the phenomena) and a strict separation of *logic from epistemology and metaphysics at the level of reason*.

Hegel subjected Kant's approach to categories and ideas to hard criticism.[8] Regarding issues treated in this chapter, the most important aspects of Hegel's critique are as follows. Kant's investigation into categories and ideas served for the purpose of discovering whether these "forms of thinking were capable of leading to the cognition of truth" (1840a, §41, Zusatz 1, 86–87; 1991, 82; 1959, 84). But, according to Hegel, Kant never investigated the inner connections of these forms in the *process of their use*, i.e., the process of cognition. In this process, they develop from one another; they show their own inherent limitations by revealing the restriction of their usage to a certain level of cognition. Hegel claims that ((1840a, §41, Zusatz 1, 87; 1991, 82; 1959, 84)

> [t]he forms of thinking are the object and the action of the object itself. They examine themselves. On themselves they have to determine their own limits and their defects (*Mangel*).

What is required is a reconstruction of the sequence and interdependence of the forms of thinking in the very process of thinking. The subject matter of philosophical investigation should be the *logical categories in use*.

The above statement by Kant – that we can think the thing-in-itself, but we cannot know it – is justified by the claim that "otherwise we should be landed in the absurd conclusion that there can be appearance without anything to appear" (Bxxvi; 1965, 27). But this means that Kant, on the one hand, grounds his approach to pure (theoretical, speculative) reason by accepting the difference between the phenomenal forms of an object and its noumenon, but, on the other hand, he does not provide any further thought-determinations – categories – which would make the category of noumenon more determined, and turn it into a cluster of categories. Such a replacement of the category of the noumenon by a cluster of categories would then, of course, require a complete restructuring of Kant's table of categories.

Hegel, in order to undertake this particular endeavor, creates a new framework for it. The framework characterizes the *logical*, *epistemic* (*cognitive*), and *ontological* status of the categories of thinking.

Hegel begins by targeting the central problem of epistemology: "[h]ow do we, as subjects, get over to the objects," (1842, §246, Zusatz, 13; 1970a, 8; 1970b, Vol. 1, 198), i.e., "the difficulty lies in the passage from the thing to the cognition which is accomplished by means of reflection (*Nachdenken*)" (1840a, xiv; 1991, 6).[9] According to Hegel, to appropriate a thing means to think it. "[I]ntelligence familiarizes itself with things . . . by thinking them, it sets their content into itself" (1842, §246, Zusatz, 16; 1970a, 9; 1970b, Vol. 1, 200); "[t]he object (*Sache*) for us can be nothing else but our notion of it" (1923, T. 1, 15; 1969, 36). By thinking *real singular objects* "we add to them, so to say, the form of universality . . . This universal of things is not a subjective belonging to us, but rather a noumenon opposed to the transitory phenomena; the true, the objectivity, the real of the things themselves" (1842, §246, Zusatz, 16–17; 1970a, 9; 1970b, Vol. 1, 200).

Therefore, in Hegel's view, human thinking has an *objective content*. One can speak about *objective thinking* (1923, T. 1, 31; 1969, 49) or about "*objective* thoughts," (1840a, §24, 45; 1991, 56; 1959, 45) because "[t]hinking [is] the activity of the universal . . . which with respect to its content exists solely in the object and its determinations . . . the thinking is with respect to its content only true insofar as it is immersed (*Vertieft*) into the object" (1840a, §23, 44; 1991, 55; 1959, 45). Hegel claims that via the phenomena, understanding is capable of penetrating into the inside (*Inneres*) of things (1928, 111; 1967, 191).

This means that the content of thinking, because it is an objective one, can serve for a "more exact determination and discovery of *objective* (*gegenständliche*) relations" (1923, T. 1, 13; 1969, 35).

And there is, according to Hegel, a certain hierarchy of the forms of human consciousness, which can be symbolized as follows (1840a, §20, 35–36; 1991, 49–50; 1959, 36–39)[10]

Sensual Feelings, Intuitions, Will, Desires, etc.
↓
Representations
↓
Thoughts
↓
Notion

He claims that "[t]he *content* which fills our consciousness, independently of what type it is, makes up the *determinations* of our feelings, intuitions, images, representations, of our purposes, duties. etc., and of our thoughts and notions" (1840a, §3, 6; 1991, 26; 1959, 6). This content is the *objective activity* of consciousness; "let it be feeling, belief, or representations – *thinking in general* has not been inactive. Its own activity and production are *present and contained in it*" (1840a, §2, 5; 1991, 25; 1959, 5). The content of our knowledge about the thing-in-itself is being produced by means of the activity of this very thinking; its content is "the *object* of consciousness" (1840a, §3, 6; 1991, 26; 1959, 6). Hegel states (1840a, §22, 42; 1991, 54; 1959, 43)

[b]y means of thinking something is *altered* in the way in which the content initially is feeling, intuition, or representation. Thus, via *change*, does the *true* nature of the object come to our consciousness.

This process of the movement of the content of our thinking has the character of cognition; "it can be only the *nature of the content* itself which moves in scientific cognition" (1923, T. 1, 6; 1969, 27).

Three consequences immediately suggest themselves from Hegel's approach to thinking. Because he claims that it is a specifically human process by means of which we create the objective content of our thinking, which has the status of cognition, Hegel *does not* – contrary to Frege (1892, 62) – *separate the subjective performance of thinking from its objective content.*

What also follows from Hegel's claim that thinking has an objective status, because it is able by its own productive (*logical*) activity to *cognize* the *thing-in-itself*, is that categories of thinking unify in themselves a *logical*, *epistemic* (*cognitive*) and *ontological* dimension. What Hegel requires from the philosophical investigation into logical categories is that (1923, T. 1, 18; 1969, 39):

the content be drawn into the thinking treatment (*denkende Betrachtung*) . . . With this introduction of the content into the logical treatment, not the *things* (*Dinge*), but the *object* (*Sache*), the *notion* of the things becomes the subject matter.

This means that Hegel *merges into a unity* the three core philosophical disciplines: logic, epistemology, and metaphysics (1923, T. 1, 46-47; 1969, 63).[11] Hegel's *Science of Logic* and *Encyclopaedia Logic* represent the realization of their unification, a trend implicitly inaugurated by Kant's first *Critique*, although, as shown above, the latter ends explicitly in their mutual separation.

Finally, Hegel's approach to logical categories enables one to understand why he introduces the "Essence"-cluster of categories. Given his view – that we can think the thing-in-itself, penetrate it by means of thinking, and so acquire knowledge about what is given beyond the level of the phenomena – he inevitably *replaces Kant's category of noumenon* by the categorial cluster "Essence." But by replacing the former by the latter, he in turn *reinterprets Kant's category of phenomenon*, so that it becomes either the categorial cluster "Appearance" ("*Schein*"), or the categorial cluster "Manifestation" ("*Erscheinung*"), depending on the *cognitive (epistemic) status of this phenomenon with respect to the knowledge of the essence*. In case when knowledge of the phenomena serves as a starting point for the production of knowledge of the essence, Hegel labels these phenomena as "appearances." In case the knowledge of the essence becomes the starting point for the production of knowledge about the phenomena, he labels the latter "manifestations." By distinguishing between the two types of knowledge of the phenomena – *one more advanced than the other* – Hegel implements his view that cognition has the character of the production of new knowledge, that our knowledge grows and develops, and that it passes through various stages. These stages are characterized by their respective categorial clusters (categorial "outfit"), where categories of a certain stage differ from those of the successive, more advanced stages. Kant's claim, that categories of understanding can be applied only to phenomena, Hegel interprets in such a way that he avoids Kant's pessimistic conclusion – that we cannot attain any knowledge of the thing in itself – and instead concludes that we can attain knowledge because (1923, T. 1, 27; 1969, 46)

> *[t]he criticism of the forms of understanding* has had as a result . . . that these forms do not *apply to things-in-themselves*. This cannot have another meaning than that, that these forms are in themselves untrue.

Categories which are necessary and sufficient for grasping the phenomena as appearances are necessary but insufficient for grasping in thought the essence-cause-ground underlying these phenomena. Thinking has to generate the latter from the former. Similarly, categories which are necessary and sufficient to grasp the essence-cause-ground are by themselves necessary but insufficient to explain why a certain essence manifests itself by means of certain phenomena.

The fact that Hegel was able to overcome Kant's separation of logic from epistemology and metaphysics, that he was able to overcome the above reconstructed five negative consequences of his *CPR*, can be understood only if one understands the *idealistic nature* of Hegel's merger of logic, epistemology, and metaphysics. Kant holds to an epistemological model, which can be expressed as follows:

Human Beings – Thoughts of Human Beings – Objects

so that "we put *the thoughts* between us and the objects as a medium (*Mitte*) in such a way that this medium separates us from the objects" (1923, T. 1, 15; 1969, 36). Hegel, on the contrary, grounds his assertion that we are connected with the object via this medium on the idealistic claim "that these very objects, which are supposed to stand beyond and at the other extreme – beyond the thoughts referring to them – are themselves thought-objects" (1923, T. 1, 15; 1969, 36).

Hegel criticizes Kant because, by postulating the thing-in-itself, "the logical determinations remained burdened with (*behaftet*) the object . . . , and a thing-in-itself, an infinite impulse (*Anstob*) was left on them as a beyond" (1923, T. 1, 32; 1969, 51). The fact that we as humans are able to penetrate in our thinking beyond the phenomenal forms of the objects and grasp their inside, is, according to Hegel, just the result of the fact that *what penetrates* – the thinking – and *what is penetrated* – the object – are, so to say, made of the same "stuff" – *thoughts*. This is readily seen from Hegel's characterization of the case when consciousness, via the play of forces, looks into the true background of things (1928, 110; 1967, 190). According to him (1928, 128; 1967, 212)

> reason experiences . . . in the inside of the appearances only *itself*. Raised above perception, consciousness reveals itself united with the supersensuous world via the medium of appearance. The two extremes, the one that of the pure inside, the other that of the inside looking into the pure inside . . . merge together.

Hegel transforms everything into the objective content of thinking – the ". . . objective thought is the inside (*Inneres*) of the world" (1840a, §24, Zusatz 1, 45; 1991, 56; 1959, 46) – while simultaneously separating the latter from human beings and designating this content as the feature, the determination of the notion. This notion, by its own operations (*logical* aspect), gradually, by passing through various stages, cognizes (*epistemic* aspect) itself, and simultaneously, via these stages, develops its own "body" (*ontological* aspect). Hegel's critique of Kant, as well as his merger of logic, epistemology and metaphysics, are based on a *thoroughly accomplished idealistic construction.*

In this subchapter I have been analyzing the *positive* aspects of Hegel's idealism by means of which he was able to overcome the negative consequences of Kant's approach to theoretical (pure, speculative) reason. Now it is necessary to examine the *negative* aspects of Hegel's idealism for the philosophical understanding of theoretical reason.

11.3 HEGEL ON ESSENCE vs. MARX ON ESSENCE

If one takes Hegel's categorial clusters of "Being" and "Essence" from his *Science of Logic* and simply compares their structure with the structure of Marx's *Capital* (first volume and the manuscripts of the second and third volume), he or she will *not* be able to find out either how Hegel's idealism penetrated into the structure of those clusters, or how Marx restructures them.

In order to approach this discovery, one should understand, first, why certain categories have their respective position in the Hegelian categorial structures. Here I mean especially the categories of *measure*, *substance*, *universal*, and *contradiction*. Secondly, one should understand the idealistic nature of the sequence of the categorial clusters of "Being," "Essence," and "Notion." Thirdly, one should understand the consequences of Hegel's idealism for the relation of philosophy to the special (natural and social) sciences. Finally, one should come to understand, by investigating Marx's *Critique of Hegel's Philosophy of Right* and "Method of Political Economy" of the *Grundrisse*, how it is that Marx restructures Hegel's categorial clusters, "Being," "Essence," and "Notion," on a realistic (materialistic) basis and how this restructuring affects the relation of philosophy to the natural and social sciences.

When one studies Hegel's understanding of how it is that the logical categories of theoretical reason function – what purpose they serve – he or she will find that, according to Hegel, they do not serve as part of the process of discovering the cause-essence-ground of an object's phenomena as appearances, but ultimately, they serve as part of the process of discovering the notion, which is the true, "absolute ground" (1923, T. 2, 213; 1969, 577) of objects outside human thinking. Wherever Hegel speaks about the penetration of objects by human thinking, he claims that thinking arrives *not at the notion of the thing-in-itself*, but ultimately at the *notion of the notion*. So, for example, he mentions the case when, at the level of understanding, we discover the various laws as the internal of the external, e.g., the law of falling stones and of the movement of the celestial bodies, and then try to unify them into one law, e.g., the law of gravity. Hegel claims about this case (1928, 115; 1967, 196)

[b]ut with this unification (*Ineinanderfallen*) the laws lose their determination. The law becomes more and more superficial. It does not in fact express the unity of *these determined* laws, but is a law which leaves out their determinations. The *one* law, which unifies in itself the law of the fall of bodies on earth and of the celestial movement, does not express them.

From this, in itself a true claim, Hegel draws the idealistic conclusion that "[t]he unification of all laws in the *universal attraction* does not express any content other than just the *bare notion* of law" (1928, 115; 1967, 196). Cognition by that synthesis does not discover yet another special law, but "only the *notion* of the law itself" (1928, 115–116; 1967, 196). The same holds also for Hegel's view on the purification of a scientific law as reconstructed in the first subchapter above. The ultimate aim of the process of purification, he argues, is (1928, 191–192; 1967, 291–292)

to purify the law and its moments into a *notion* . . . to raise the law completely into the shape (Gestalt) of the notion and to detach all *its moments from a determinate being* . . . The result of such experiments cancels in such a way the moments or the inner specifications (Begeistungen) as properties of the determined things and makes the predicates free of the subjects.

An idealistic framework also determines Hegel's approach to the categories of the concrete universal, substance-subject, and contradiction.

In the case of the category of a universal, Hegel distinguishes between two types of knowledge of the universal: that of the *concrete universal* and that of the *abstract universal*. The abstract or sensuous universal is a result of the understanding which puts together sensuous singulars according to certain features or signs common to them. These features, signs, and the singulars themselves are not unified by that universal into an order by necessity, but only for the purpose of attaining an overview. The content and activity is not in the universal, but outside it, in the singulars which are split up (*zersplittert*), dismembered (*zerstückelt*) and isolated (*vereinzelt*) in the universal (1842, §246, Zusatz, 17–19; 1970a, 10; 1970b, Vol. 1, 201; 1928, 89–91, 185–186; 1967, 162–163, 284–285; 1923, T. 2, 263, 291; 1969, 621, 647). If we are to understand a universal as a concrete universal, this means that it should display in itself the feature of a necessary relation between the singulars, the characteristic of an internal unrest (*Unruhe*) (1928, 208; 1967, 311–312), as well as diversity and manifoldness (1923, T. 2, 242, 293-294; 1969, 602, 605). The knowledge of the universal as an abstract, sensuous composite should be surpassed and overcome by a knowledge of the universal as a concrete one (1842, §246, Zusatz, 20; 1970a, 12; 1970b, Vol. 1, 203)

[i]f the genera and forces are the inside of nature and, against them as the universal, the outside and the singular is transient, then one requires as a third stage the internal of the internal, which . . . would be the unity of the universal and particular.

What Hegel also correctly, at least in my view, requires is that scientific laws should acquire the status of a concrete universal (1928, 208; 1967, 311–312). But, on the other hand, Hegel views the concrete universal only as the feature of the *spirit, notion* (1840a, §21, Zusatz, 43; §24, 46; 1991, 5, 56; 1959, 43, 46) and not as a feature of the objects outside human thinking, because "things are singular and the lion in general (*überhaupt*) does not exist" (1842, §246, Zusatz, 13; 1970a, 7; 1970b, Vol. 1, 198). Therefore, Hegel labels this understanding of such type of universal as the *recognition of the notion* (*begreiffendes Erkennen*) (1842, §246, Zusatz, 21; 1970a, 12; 1970b, Vol. 1, 203).

Hegel proceeds in a similar way when discussing the case of the category of substance. Its introduction into the categorial subcluster "Reality" is based on his reflections from previous subclusters, where he dealt with issues like the relation of the inside of things to their outside (external manifestations); the relation of the essential to the non-essential; the unification of different phenomena into one scientific law; and the introduction of the category of the substrate, which should be a unity of all apparent qualities.[12] Unfortunately, according to Hegel, everywhere throughout this territory one faces unresolved problems. A person deals here not with self-relation but only with relations; the essential is not properly understood because it is essential only with respect to the non-essential; the law only has the character of a "*quiet* copy of the world" (1923, T. 2, 127; 1969, 503); the substrate is just a mere "also" ("auch") of apparent qualities (1923, T. 2, 118; 1969, 495) and forfeits its independence from them, though it ought to have an independence somehow related to the essence. In order to solve all these problems Hegel introduces the category of substance. This category should lead cognition to the totality proper: to the understanding of self-relation and self-movement. Here Hegel also introduces the category of causation, finally, to arrive at its characterization as *mutual interaction* (*Wechselwirkung*).

Again, on the one hand, Hegel correctly stresses that in order to solve the problems noted above one has to broaden the category of causation so that it encompasses not only unidirectional causal action but also mutual action, mutual production and reproduction, i.e., he acknowledges the need to arrive at the category of *causa sui*. But, on the other hand, Hegel understands this broadening of the categorial apparatus as a development of cognition from the cluster "Essence" to that of "Notion." The category of mutual interaction should enable one to understand that "[t]he cause meets . . . simply and solely itself" (1923, T. 2, 216; 1969, 579). Hegel views this as the ultimate stage of the development of the category of substance in the cluster "Essence." As he states, this is its "consummation . . . But this consummation is no longer the *substance* itself, but something higher, the notion, the *subject*" (1923, T. 2, 216; 1969, 580). With respect to the category of substance, causality, and mutual interaction Hegel makes the explicit idealistic claim that "[t]he *dialectical movement* of the substance through causality and mutual interaction is . . . the immediate genesis of the *notion* by means of which its *becoming* is expressed" (1923, T. 2, 214; 1969, 577) and (1923, T. 2, 219; 1969, 582)

> [t]he mutual opacity of the substances standing in the causal relation has disappeared because the originality of their self-subsistence has passed into a positedness and, therefore, into a self-transparent *clarity*. The original object is *this*, in that it is only the *cause of itself*, and that is the *substance liberated into the notion*.

Even if Hegel treats the category of substance both as a part of the cluster "Essence" and the cluster "Notion," the category does change depending on which cluster subsumes it, for when it is part of essence, contrary to notion, substance *lacks the character of a self-related, self-moving, internally differentiated entity*. Therefore, Hegel, at the level of the judgment of necessity, claims that (1923, T. 2, 293–294; 1969, 650)

> [t]he determination to which universality has advanced itself is . . . the *in and for itself* or *objective* universal, to which in the sphere of essence corresponds *substantiality*. It is distinguished from the latter in that it belongs to the *notion* and . . . therefore . . . the *difference* is immanent to it, whereas the substance has its difference only in its accidents, but not as a principle within itself.

The reconstruction of Hegel's approach to the category of substance in the cluster "Essence" and to the category of substance-subject in the cluster "Notion" enables me to bring to a conclusion my analysis of Hegel's approach to the category of measure from chapter six, and to comprehend and explain why Hegel in the *Science of Logic* does not provide any characterization of the category of immanent measure. Because of Hegel's idealism, there exists, with respect to the clusters of "Being," "Essence," and "Notion" a *fundamental trichotomy*. Wherever (cluster "Being") he mentions the category of immanent measure, he relates it neither to the category of essence-ground-cause (cluster "Essence") nor to the category of substance-subject (cluster "Notion"). Wherever Hegel deals with the essence as essence (subcluster "Essence as Reflection in Itself"), he relates it neither to the category of immanent measure nor to that of the substance-subject. Finally, when he treats substance-subject as something internally active, differentiated, self-related, and self-moving, categories like measure and essence are viewed by him, with respect to the cluster "Notion," as already overcomed, i.e., as untrue forms of knowledge.[13]

It is exactly with respect to the issue of internal activity that Hegel brings up the category of contradiction. He deals with it in the subcluster "Essence as Reflection in Itself" of the cluster "Essence," where he claims that (1923, T. 2, 58-61; 1969, 439–442)

> [t]he contradiction . . . is the root of all movement and vitality (*Lebendigkeit*). Only insofar as something has a contradiction within itself does it move, have an urge (*Trieb*) and activity . . . Something is . . . alive only insofar as it contains contradiction within it . . . Only when being driven to the highest point (*Spitze*) of the contradiction do the manifolds become mutually active and vivid (*lebendig*), and acquire in it . . . the indwelling pulsation of self-movement and vitality.

But, again, because of his idealistic starting points, Hegel – with respect to the reflection-determination of essence – claims that "[t]he notion is the *concrete* and *richest* because it is the ground and totality of the preceding determinations of the categories of being and of the determinations of reflection" (1923, T. 2, 258; 1969, 617). Therefore, the entity, which is the result of the mutual relation of the contradictory determinations, is in fact "the product of reflection of their own notion" (1923, T. 2, 494; 1969, 833). Because of Hegel's idealism the categories of the *Science of Logic* should serve *not to provide human thinking with forms which would allow us to think the thing-in-itself as internally differentiated, internally active, self-moving, and self-developing, but to think it as a moment, feature, or aspect of the self-development of the notion.*

It should be apparent now that the idealistic starting points of Hegel's philosophy determine not only its overall structure, its "macroscopic" division into *Speculative Logic*, *Philosophy of Nature*, and *Philosophy of Spirit*, and create not only the general framework of his *Science of Logic* and *Encyclopaedia Logic*, but also penetrate into their very categorial clusters, and determine, so to speak, their "microstructure," i.e., the respective position of individual categories as well as the location of the categorial subclusters. Therefore, any attempt at a comparison of the logical structure of Marx's *Capital* with that of Hegel's *Science of Logic* and *Encyclopaedia Logic* has to accommodate the fact that *a switch from idealistic starting points to realistic (materialistic) starting points, as accomplished by Marx, inevitably leads to a restructuring of Hegel's categorial clusters of "Being," "Essence," and "Notion."*

The characteristics of Hegel's approach to the categories of concrete universal, substance-subject and contradiction, as they are reconstructed above, constitute part of an idealistically constructed developmental ontology, which can be generally characterized as a monism of the notion. Hegel views the whole of the universe as passing through three distinctive developmental stages: the development of notion, the development of nature, and, finally, the development of the subjective (human) spirit. The stage of the notion in itself passes through the stages of the objective notion and the subjective notion. The latter, at its highest level of its self-development – the absolute idea – becomes the starting point for nature and human beings. What has to be emphasized here is that both nature and human spirit – according to Hegel – emanate from the idea not in such a way "as if from outside an alien content would be added to it, but in such a way that it is the idea's own action to determine and unfold itself into nature and spirit" (1840a, §43, Zusatz, 95; 1991, 87; 1959, 91).

Hegel claims also that (1842, §249, 32; 1970a, 20; 1970b, Vol. 1, 212)

> [n]ature is to be regarded as a *system of stages*, which emerge (*hervorgehen*) one from another by necessity . . . but not in such a way as if one would be produced from the other *naturally*, but within the inner idea representing the ground of nature. The *metamorphoses* pertains only (*kommt zu*) to the notion, because only its change is development.

So, even if, according to him, nature and human spirit pass through certain developmental stages, their development represents just the development of the idea; the source of the former is located neither in nature nor in the human spirit.

This ontology, because it is just a world*view*, cannot explain by itself why Hegel's philosophy fell precipitously out of favor. To understand this, one has to take into account another feature of Hegel's philosophy, namely that which pertains to the *division of work and competence between philosophy and special (natural and social) sciences*. According to him, it is the exclusive competence of philosophy, and of course of his philosophy, to deal with the notion, which displays the characteristic of a substance-subject and concrete universal. In the case of special science, e.g., physics, holds that "[t]he universal of physics is abstract, or only formal; it has no determinations on itself, and does not pass over into the particular" (1842, §246, Zusatz, 19; 1970a, 11; 1970b, Vol. 1, 202). On the contrary, "the determinations are not indifferent (*gleichgültig*) for the universal of philosophy. It contains the difference in itself" (1842, §246, Zusatz, 20; 1970a, 12; 1970b, Vol. 1, 203). Therefore (1842, §246, Zusatz, 18; 1970a, 10; 1970b, Vol. 1, 201),

> [t]he philosophy of nature takes up the material, which physics prepared for it from experience, at that point to which physics has brought it, and transforms it . . . physics has to work into the hands of philosophy, so that the latter may translate the former's universal of understanding into the notion.

Even if both natural science and philosophy of nature deal with a universal, still the universal of the latter is superior to that of the former, according to Hegel, because it accesses it "in its *own immanent necessity* in accordance with the self-determination of the notion" (1842, §246, 11; 1970a, 6; 1970b, Vol. 1, 197). So, Hegel strictly differentiates between the cognitive competence of philosophy and that of special sciences, granting precedence to philosophy. While philosophy deals with the *notion*, the special sciences "have to bring the universal down to the empirically singular and real." Therefore "[i]n these sphere of change and chance not the *notion*, but only the *grounds* can be made to count" (1840a, §16, 24; 1991, 40; 1959, 26).

Two consequences immediately suggest themselves. First, philosophy is *superior* to all special (natural and social) science, because only through it can the true nature of everything – the notion – be recognized. To be more precise, Hegel's designates his own philosophy as the ultimate form of knowledge because it reveals to us the true nature of the universe – the notion – by looking into the inside of the inside.

Secondly, and here we find the negative consequences of Hegel's idealism, the history of special sciences and philosophy comes to an end in Hegel's philosophy. Because Hegel claims that his work reveals to us the true nature of everything – the notion – and that this is the absolute ground of everything, the cognition made possible by the special sciences comes to an end. Whatever will be discovered by the special sciences in the *future*, we know already *now* that its true nature will be the notion. So, e.g., not only after Newton has unified the celestial physics of Galileo and the terrestrial physics of Kepler in the concept of gravity, does Hegel declare that the ground of the force itself (its true inside) is the notion, but even with respect to the future – say the attempts at a conceptual synthesis of the quantum field theory and the theory of gravitation – do we *know in advance* that the true, absolute ground of this future discovery, whatever will be discovered, is also the notion!

The special sciences, with respect to Hegel's philosophy, *completely lose their autonomy*, and the facts they discover just "serve for the presentation and imitation of the original and completely independent activity of thinking" (1840a, §12, 21; 1991, 37; 1959, 22).

It is exactly these negative consequences of Hegel's idealism which were criticized by Marx in the "Critique of Hegel's Philosophy of Right." Its most important aspects, related to this chapter, are as follows.

1. In Hegel's *Philosophy of Right* (1982, 8; Marx and Engels 1975ff., Vol. 3, 8)

> [r]eality is not expressed as it is in itself, but as another reality. The ordinary empirical experience (Empirie) does not have its own spirit but an alien spirit as its law . . .

2. One faces here (1982, 8, Marx and Engels 1975ff., Vol. 3, 8)

> a twofold history, an esoteric and an exoteric. The content lies in the exoteric part. The interest of the esoteric is always that of recovering the history of the logical notion in the state [as the exoteric]. But the real development goes on the exoteric side.

3. In Hegel's approach (1982, 9; Marx and Engels 1975ff., Vol. 3, 9)

> [t]he real becomes the phenomenon, but the idea has no other content than this phenomenon. Moreover, the idea has no other logical purpose, namely to become for itself an "infinite real spirit."

4. One faces here (1982, 11; Marx and Engels 1975ff., Vol. 3, 10)

> the same transition, which in the *Logic* accomplishes (*bewerkstelligen*) the passage from the sphere of Essence into that of Notion. The same passage is accomplished in the *Philosophy of Nature* from inorganic nature into life. The same categories provide the same soul (*Seele*) now for this, now for that sphere.

5. (1982, 11; Marx and Engels 1975ff., Vol. 3, 11)

> It is important that Hegel makes the idea everywhere the subject, and turns the proper and actual subject . . . into the predicate.

Marx further criticizes Hegel because he "does not develop his thinking out of the thing (*Gegenstand*), but the thing according to a thinking which is ready-made and has become in the abstract sphere of logic" (1982, 15; Marx and Engels 1975ff., Vol. 3, 14). Even if Hegel deals in his "Philosophy of Right" with a specific subject matter, still for him "the true interest is not the philosophy of right, but logic . . . Logic is not used to prove the state, but the state is used to prove logic" (1982, 18; Marx and Engels 1975ff., Vol. 3, 17–18). Marx opposes Hegel's separation of properties ("predicates") from the real thing ("subject") by making the realistic claim that in scientific cognition one has to start from the real thing, "*ens* (*ὑποκείμενον*, subject) . . . [to] consider the universal . . . to be the actual essence of the real finite, i.e., existing, finite or the real *ens* as the true subject of the infinite" (1982, 25; 1975ff., Vol. 3, 23–24). This claim of Marx's in turn leads him to the epistemological claim that the aim of a special science, targeting a concrete ens, is not – as it claimed by Hegel – to "find everywhere the determinations of the logical notion, but to grasp the specific logic of the specific thing" (1982, 100; 1975ff., Vol. 3, 91). This means that according to Marx different kinds of objects have different grounds-essences-causes, and, therefore, *the investigation into them belongs exclusively to the competence of the respective special (natural and social) sciences*. In this sense I claim that Marx, via his realistic (materialistic) critique of Hegel, reestablishes, *vis-à-vis* Hegel's philosophy, the autonomy of the special sciences.

Marx's realism, as opposed to Hegel's idealism, is also a part of his logical and epistemological considerations which create the framework of his economical works; here I mean primarily his "The Method of Political Economy"[14] from the manuscript *Grundrisse*. In this discussion, Marx, contrary to Hegel, insists on the difference between the real thing and the thought-thing, and insists that the development of the objective content of thinking via a method of advancement from the abstract to the concrete does not abolish the real thing (*das Subject*). He states this explicitly as follows (1976, 36-37; 1975ff., Vol. 28, 38)

[t]he concrete is . . . the real starting point and, therefore, also the starting point for intuition and representation . . . the method of advancement from the abstract to the concrete is only the way in which thinking assimilates the concrete and reproduces it as thought-concrete . . . [i]t is in fact a product of thinking, of comprehension (*Begreiffen*) . . . of the transformation of intuitions and representations into notions. The whole as it appears in the head (*Kopf*) as a thought-whole is a product of the thinking head . . . The real thing (*Subject*) still remains to exist in its independence outside the head as long as the head behaves only speculatively, theoretically.

This means that for Marx categories of theoretical reason are categories of conscious reflection, while the method of advancement from the abstract to the concrete is the method of the creation of this reflection by the very productive activity of thinking.

The fact that Marx has a realistic understanding of what the thought-object (thought-concrete, thought-whole) stands for, enables us to understand why Marx in his works on economics explicitly mentions two types of thought-object depending on what scientific explanation aims at. In the case when one targets just one object of a kind, so that it stands for all its objects and, therefore, also for the kind as such, and is interested solely in the theoretical reconstruction of its internal structure, then he or she can proceed by the method of scientific explanation based on the movement from the thought-abstract to the thought-concrete. This type of scientific explanation represents thought-operations with an object labeled by Marx as "universal type" ("*allgemeiner Typus*") (1992, 215; 1967, Vol. 3, 143) or as an object "in its ideal average" (1992, 853; 1967, Vol. 3, 831), and for which I will use the term "ideal average type of thought-object." Knowledge produced by means of operations with such a type of thought-object does not explain the specific differences between the individual objects of the same kind. What is also left unexplained are the specific historic trajectories in which their respective internal structure originated and which are somehow preserved in them. Knowledge generated by means of thought-operations with the ideal-average-type of an object serves with respect to those individual objects as a theoretical mean to determine their principal, universal determinations. According to Marx (1992, 215; 1967, Vol. 3, 142–143)

[i]n such a general investigation it is usually always assumed that the real conditions correspond to their concept, or, what is the same, that the real conditions are represented only in so far as they express their own universal type.

With respect to the capitalist mode of production this means, according to Marx, that "we need to present only the inner organization of the capitalist mode of production, in its ideal average, as it were" (1992, 853; 1967, Vol. 3, 831). Of course it is the task of the theoretician to find out among the individual objects of the kind under investigation one whose internal structure is not distorted by remnants of its own historical past when it was as yet a (different) object with a different essence.

Marx, for example, in his investigation into capitalistic industrial production, chose as such a type of object the industry of England.

If one aims at the investigation into and explanation of the particular characteristics of the individual objects of the kind, then he or she has, *after* acquiring the knowledge about their universal determinations, to view the individual object – initially chosen as the ideal-average-type – just as an *original* type, while the other individual objects different from it are original types in their own right. The task of scientific explanation based on the original type of a thought-object is to derive those characteristics of individual objects which make them different from that original type which initially served as the ideal-average-type of an object.

Marx used the notion of the original type in his reflections about the indigenous (*naturwüchsig*) communal property. He claimed that (1980a, 113; Marx and Engels 1975ff., Vol. 29, 275)

> [i]t is an early form which can be found among the Romans, Teutons and Celts, and of which a whole sample-card (though only as remnants) is still in existence among the Indians. A more careful study of Asiatic, particularly Indian, forms of communal property would prove how the various forms of indigenous communal property give rise to various forms of its disintegration. For instance, the various original types of Roman and Teutonic private property can be derived from the various forms of Indian communal property.

What is readily seen here is that the *knowledge* about the various forms of communal property in India, i.e., the thought-representation of real objects serves as a point of departure for the production of new knowledge, new notions, which express the nature of the various forms of Roman and Teutonic private property, i.e., it represents thought-objects which after being subjected to thought-transformations yield features corresponding to these forms of private property.

Marx's realism virtually cancels out Hegel's *Philosophy of Nature* and *Philosophy of Spirit*, but Marx picks up on Hegel's speculative logic (as explicated in *Science of Logic* and *Encyclopaedia Logic*) and fundamentally restructures it. One of central aspects of this process involves Marx's criticism of Hegel's approach to the category of contradiction. While Hegel uses the symbols $+A$ and $-A$ to express his understanding of contradiction,[15] Marx strictly distinguishes between two cases of contradiction. One, labeled by Marx as "real extremes" ("*wirkliche Extreme*") (1982, 97; Marx and Engels 1975ff., Vol. 3, 88) is typical for the case when we have *two different essences*. They do not require each other; they do not complement each other. As an example Marx uses the case of pole and *non*-pole, human and *non*-human species (1982, 98; 1975ff., Vol. 3, 88). The second type of contradiction Marx comprehends as opposed (*entgegengesetzte*) determinations of *one* essence, or the difference inherent in *one* essence at its highest stage of development. This type of contradiction belongs to a differentiated essence; "they are what they are *only* as a differentiated determination and as this *differentiated* determination of essence" (1982, 98; Marx and Engels 1975ff., Vol. 3, 88). For this case Marx uses the example of the mutual attraction between the north and south poles and the mutual attraction between male and female sexes. Their mutual attraction Marx understands on the basis of the fact that "their essence is identical" (1982, 98; Marx and Engels 1975ff.,

Vol. 3, 88). Marx views the approach to the category of dialectical contradiction via the category of real extremes as a specifically Hegelian approach and states that (182, 97–98; Marx and Engels 1975ff., Vol. 3, 88)

. . . when Hegel treats universality and singularity, the abstract moments of the syllogism, as real extremes, then here is exactly this basic dualism of his logic. The further development of this [question] belongs in the critique of Hegelian logic . . .

This means that Marx's understanding of contradiction differs from that of Hegel, who understands it as a relation of real extremes: pole and *non*-pole; human species and *non*-human species. Marx's understanding of the category of contradiction cannot, therefore, be symbolically expressed by means of a predicate-variable A as $+A$ and $-A$; "[t]here is no real dualism of essence" (1982, 99; Marx and Engels 1975ff., Vol. 3, 89). Marx also claims that (1982, 100; Marx and Engels 1975ff., Vol. 3, 91)

Hegel's chief mistake consists in that he conceives the contradiction in manifestation as *unity in essence*, in the idea; whereas it certainly has something more profound for its essence, namely, an *essential contradiction.*

The fact that Marx here understands his own approach to the category of contradiction, as one that is opposed to Hegel's identifying approach, enables me to relate Marx's ideas on production and consumption from the *Grundrisse* to his views on the category of contradiction, because in the introduction to this work he claims about that relation that "nothing is simpler for a Hegelian than to posit production and consumption as identical" (1976, 30; Marx and Engels 1975ff., Vol. 28, 31). By examining Marx's "anti-Hegelian" approach to the relation of production and consumption, it is possible to reconstruct three principal points that mark his understanding of the extremes of an essence (1976, 29–30; Marx and Engels 1975ff., Vol. 28, 30–31).

1. *Immediate identity of the opposites.* Each contains in itself the moment of the other.
2. *Each of the opposites is a mean for the other and is mediated with itself via the other.* Their mutual relation of dependence has the character of a movement in which one is indispensable for the other.
3. *The opposites of an essence mutually produce and reproduce each other.* Once the unity of the opposites comes into being, it is continuously produced and reproduced by their mutual action; "[e]ach of them is not only directly the other, nor does it merely mediate the other, but each of the two, by the fact of its taking place, creates the other, creates itself as the other" (1976, 31; Marx and Engels 1975ff., Vol. 28, 31)

In this way, Marx replaces Hegel's identity in the idea with the relation of opposites in the essence, as a result of which, "essence" as defined by Marx *moves*, i.e., is an *active entity*. This means that in Marx's approach, essence is thought of as *active*, *self-*

moving; it is thought of as *differentiated in principle*. This is the *first* central aspect of Marx's realistic (materialistic) restructuring of Hegel's categorial clusters of "Being," "Essence," and "Notion." Because, as shown above, Marx takes the *universal to be the essence of the thing*, its *second* aspect is that he understands not the notion of a thing (from which the latter should somehow emanate), but its *essence as a concrete universal*. This means that Marx shifts the category of the concrete universal from Hegel's cluster of "Notion" and puts it into the subcluster "Essence as Reflection in Itself." Finally, because Marx, as shown in chapters five and six, deals with the substance of value and surplus-value, and views the substance of value as internally differentiated (concrete labor/abstract labor), he also shifts the category of the substance-subject from the cluster "Notion" and locates it directly into the subcluster "Essence as Reflection in Itself." This is the *third* central aspect of Marx's approach. By means of these three aspects it is possible to understand why Marx in his economical works, as shown in chapter six above, transforms Hegel's claim "[t]he essence has to manifest itself" (1923, T. 2, 101; 1969, 479) into "the immanent measure has to manifest itself." Because Marx places the category of substance-subject into the Hegelian subcluster "Essence as the Reflection in Itself," and views this substance both as the quantitative and qualitative source of the production of essence-ground-cause and as internally differentiated (differentiated in its quality), he is able to understand that what causes the essence to manifest itself by necessity is its own immanent measure.

After having analyzed this threefold restructuring of Hegel's categorial clusters by Marx, by taking into account my analysis of *Capital* from the previous chapters, and by taking simultaneously into account Marx's (1977) précis of Hegel's *Minor Logic*, where he states his view on the categories quality, quantity, and measure from the Hegelian cluster "Being" as follows (1977, 428)

> [t]he sphere of Being corresponds to the sensuous consciousness. They are the forms of thinking corresponding to the sensuous consciousness [,]

I can posit the full results of the investigation as follows. Hegel's cluster "Being" is restructured in such a way that it becomes the cluster "Apparent Being" and is set up by the following three subclusters:

1. The Apparent Quality
2. The Apparent Quantity
3. The Apparent (External) Measure and the Formal Ground.

Hegel's cluster "Essence" with the subclusters "Essence as Reflection in Itself," "Manifestation," and "Reality" is restructured in such a way that it becomes the cluster "Organic System" with the following three subclusters:

1 The Essence-Producing Process of an Organic System
2. The Process of Essential Formality and its Place in an Organic System
3. The Organic System as a Whole

The first subcluster here is created by the following categorial sub-subclusters: the real ground; the essential conditions; the immanent measure; the quantitative and qualitative determinations of the essence-producing substance; this substance as internally differentiated (substance-subject and concrete universal); the qualitative and quantitative determinations of the substance as the essence of quality and quantity; and the inherent types of scientific law and scientific explanation based on them.

The second subcluster contains the following categorial sub-subclusters: the life-process of a local organic system; causal loops, the relation of the local and the global process in organic system; part and whole; and the internal and the external.

Finally, the third subcluster is composed of the following categorial sub-subclusters: the genesis of the manifestation of the immanent measure (the relative and equivalent form); the forms of manifestation of the essence; modification conditions; the scientific inherent idealizational type of scientific law and scientific explanation based on it; the quantitative deviations of the phenomena as manifestations from the essence-ground-cause; explanation and critique of false and transposed consciousness.

These two categorial clusters, together with their subclusters and sub-sub-clusters stand *only for categories of theoretical reason*, i.e., *only for forms of conscious reflection*. In this book, I did not investigate the categorial cluster which would contain categories for practico-spiritual activity, like that of a norm, of an ideal, that of good and of bad, of beautiful and of ugly. These categories in their unity are involved in the creation of our projects of (social and natural) reality. This book is, therefore, only an investigation of theoretical reason.

CONCLUSION

My approach to the issue of scientific laws and scientific explanation was based on an analysis of laws which are formulated in two different spheres of knowledge: classical mechanics and Marx's political economy, each of them dealing with a specific sphere: the former with nature, the latter with society. As a consequence of this approach, differences between laws of nature and laws of society were left unanalyzed. Also left unanalyzed were the differences between the epistemic, cognitive approaches of natural science and those of social science.

The result of my critique of the Poznań school as given in chapters three, five, and six was that science can and in fact does formulate more advanced types of scientific laws as compared to the pure idealizational type of scientific law. This result pertains to *science in general*; it is, so to say, part of the *general philosophy of science*. Because I spoke about science in general, I left also unanalyzed and uncriticized one aspect of Leszek Nowak's approach to philosophy of science, namely, his assumption that both natural and social sciences use the same methods. Having claimed that Marx is the Galileo of social science, L. Nowak neglected to reconstruct those features of Marx's method which make it different from the methods applied by classical mechanics.

For an understanding of their differences, let me start from a comparison of laws of nature and laws of society. What makes them similar, at least in my view, is that both are a form of (non-simple) reproduction of their own essential (inherent) conditions. But there exists one fundamental difference between them, namely, their relation to human beings. Let us take laws of nature first. They regulate the relation between "things," e.g., the relation between the moon and the earth, or between aspects, features of a "thing." These relations existed prior to the time when they became a form of human thinking and a form of practical activity, e.g., in cosmological speculations and in space flights. *To exist laws of nature need not become a form of human action.*

Contrary to this, laws of society are forms of cooperated practical human action. Let me take, as an example of this, the law of surplus-value. Its inherent conditions represent certain specific relational characteristics of human beings which act as suppositions which, in turn, generate certain relations between human beings engaged in the process of their mutual action, and, therefore, also generate the creation of certain characteristics of these beings. And because of this action, these suppositions become the result of this action. *Laws of society can exist only as forms of cooperated practical human (inter)action.* But the latter is always an aim-setting form of (inter)action. Before cooperated practical human action starts, its aims, its goals exist in an ideal form in the spiritual activity of those who participate in this action. These aims need not incorporate the knowledge of the laws which regulate this activity. On the contrary, they can, and very often do, constitute a limited, one-sided, fragmentary view what is really going on in that action. In the case of the law of surplus-value, this is readily seen in the fact that surplus-value is generated from surplus-labor only if the owner of the labor power interacts with the owner of the means of life and production. But they enter into mutual interaction only because the former anticipates in his mind the acquiring of a *wage*, while the latter enters into this relation with the aim to obtain a *profit*. And neither of them needs to know that the wage is somehow related to the variable capital or that industrial profit has its source in surplus-value. The consciousness of actors of the interaction can be fixed at the level of the knowledge of social appearances. This form of consciousness I label "preconceptual understanding" (*vorbegriffliches Verstehen*) and it represents a real (ontical) presupposition for the existence of social laws.

Laws of society for their very real, objective existence require the existence of aims in the mind of the actors of practical activity, even if, very often, these aims have the character of a false, transposed, illusionary knowledge about what is really going on in that action. Compared to this, the situation with laws of nature is completely different. For example, in the case of the moon-earth relation, laws of nature determining that relation existed prior to the various illusionary forms of knowledge human beings had about this relation.

Because I understand real laws (of nature and society) as forms of (non-simple) reproduction, and because I view laws of society as a form of practical human action, I claim that *social laws are invariant forms* – given certain social conditions – *of practical human (inter)action*. The invariant forms of practical human action (e.g., the production of surplus-value) and the existence of aims (the aim to acquire wage and profit), even if they are different, cannot be separated.

These differences between laws of nature and laws of society have profound implications for the difference between methods applied by natural science and those applied by social science. As shown in previous chapters, in natural science cognition should proceed from appearances *via* essence to manifestations. Because the object of cognition of natural science exists prior to our, very often illusionary, empirical and theoretical cognition, it can be approached in our cognition *as a system exclusively from the outside*.

Social science should, like natural science, proceed from appearances *via* essence to manifestations. But in this sequence another aspect, completely absent in natural science, comes in. As the very ground-essence regulating certain aspects of human interaction itself has the character of human interaction, it is inseparably interwoven with an aim-setting consciousness, the existence of aims in mind. This means that while in theoretical natural science false consciousness appears, at most, at the level of phenomena (appearances and manifestations), in theoretical social science it has to be present also in the reconstruction of the very essence-cause-ground of these phenomena. *Therefore, the investigation in social relations of a certain historical period requires that the social scientist includes into his theory as a whole (i.e., at the level of appearances, essence, and manifestations) the reconstruction of aims given in mind, which very often have the character of an illusionary, false form of knowledge about what is really going on in the interactions constituting social relations.* I label this reconstruction "conceptual understanding" (*begreifendes Verstehen*) and it represents for the social scientist a special way of entering into and grasping of the preconceptual understanding of actors of actions.

The fact that the theoretical formulation of laws in social science should involve not only a grasping of what is objectively going on in human action, but should also involve the reconstruction of the various forms of (illusionary) consciousness interwoven with this action, implies that in *social science the view from the outside of the object as a system under investigation is unified with the view from the inside of the object, with the subjective views of the actors of practical action about their own action*. It remains the task of a future investigation to find out in which way it becomes possible to shift between these two perspectives (the perspective of the participant of interaction and of the "observing" theoretician). But still the following can already be stated on the relation between these two perspectives. The process of conceptual understanding in itself consists of two phases: one *empirical* and one *theoretical*. The first represents an *empirical* way of entering into the preconceptual understanding of the actors about their own action. Then the second phase sets in, where a *theoretical* grasping of the process of production and reproduction of a historically specific type of society has to reconstruct *how this (false) consciousness is produced and reproduced in the course of human interaction*. This phase, I claim, has the character of the view from the outside, and should contain a explanation of those forms of (false) consciousness, which were initially grasped by the first (empirical) phase of conceptual understanding. This explanation should have the character of a thought derivation of the sense of those terms, by means of which actors as laymen name and interpret their own life-process. While the chapters of this book have attempted to show that for the construction of the view from the outside one can utilize categories of causality, it again remains the task of a future investigation to find out what categories can be used for the first (empirical) phase of conceptual understanding.

As an example let us look into Marx's unification of the approach to the social object from the outside with the approach from the inside in his grasping of the law of surplus-value. First, at the level of appearance, in Part II of the first volume of Capital, labeled *Transformation of Money into Capital* (1990, 127–153; 1967, Vol. 1, 146–176; Marx and Engels 1975ff., Vol. 35, 157–186), he brings in a new type of aim. It is that of making money. He expresses it by the formula money $\rightarrow$ commodity $\rightarrow$ money$'$, $M \rightarrow C \rightarrow M'$, where $\Delta M = M' - M$ stands for this profit. This formula expresses the form of consciousness of the industrial capitalist, which Marx characterizes as "[t]he restless never-ending process of profit making alone" (1990, 133; 1967, Vol. 1, 152–153; Marx and Engels 1975ff., Vol. 35, 164). Second, at the level of essence, Marx in his theoretical reconstruction of the law of surplus-value (manuscript *Grundrisse* and the manuscript 1861/63) explicitly incorporates the aim of making money (profit) into the formulation of its essential conditions. Third, at the level of manifestation, he then explains, in the manuscript of the third volume of *Capital*, how this form of false consciousness is produced and reproduced by the intersection of the law of surplus-value – as a form of human (inter)action – with other forms of human (inter)action: economic competition, money-lending, lease of land, etc.

Because Marx utilizes in his construction of *Capital* a method which differs in certain aspects from that applied in classical mechanics, and in natural science in general, this is yet another proof that Marx is not the Galileo of social science, as

L. Nowak has claimed. Despite this, as I tried to show in previous chapters, and as I claim in this conclusion, *society can become subject to objective procedures of scientific research and explanation, even if the latter incorporate certain aspects which are completely absent in the methods of research and explanation used by natural science.*

NOTES

INTRODUCTION

1 To the best of my knowledge only A. Doz's (1974) paper, H. Metzler's (1977) paper, A. von Pechmann's (1980) book, V. Černík's (1986) book, and Ph. Mongin's (1989) paper deal with the category of measure.

CHAPTER 1

1 This is the standpoint of P. Achinstein (1981, 218), for example.
2 The ontological basis of the D-N model is only outlined here. Its detailed reconstruction will be undertaken in Chapter 2.
3 I will limit myself to the reduction of the law expressed by formula (1.4) to be arrived at by dropping the concept of force. To obtain a bona fide historical form of the law of pendulum one ought to study above all the writings of G. Galilei and Ch. Huygens (on this see also Sarlemijn 1993).
4 At this point I will again confine myself to the reduction of the scientific law expressed by formula (1.7). A reconstruction of the historical form of the law of fall is given, e.g., in (Renn et al. 1998).

CHAPTER 2

1 Although Popper does not investigate this case, I will show below that a statement expressing it is a scientific law.
2 Explicitly it became the subject matter of a special investigation for first time, to best of my knowledge, in U. Will's (1984) paper.
3 *Tractatus 6.3667.*
4 "$a \rightarrow b$" is the name for an ordinal conditional with *a* as an antecedent and *b* as an consequent.
5 Like Popper, Nerlich and Suchting do not scrutinize the structure of the statement *L*.
6 I analyze neither Popper's article (1967) nor his 1968 addendum to Chapter X*, where he gives up the attempt to provide a definition of natural necessity. I also do not analyze Suchting's response to them in (1969).
7 Introduced by J. S. Mill in (1950).
8 A strict conceptual differentiation of these two types of conditions will be given in Part II, subchapter 4.1, in connection with the reconstruction of the theoretical causal law.

CHAPTER 3

1 This problem will be scrutinized in subchapters 4.2 and then in chapter eight, where I analyze the relation of a natural kind to its own essence.
2 P. Murray came in (1990) very close to the reconstruction of this distinction. He claimed that "[u]nder [the] . . . dialectical (or internal logical) conception of essence and appearance, science is no longer a one-way street that externally relates appearances to the essence, but works both from the appearances to the essence *and* from the essence to the appearances" (1990, 133–134). Surprisingly, he did not realize that Marx uses two different terms, "*Schein*" and "*Erscheinung*" for this case of scientific proceeding.
3 Hegel's thoughts, 160 years before Mackie, on the purification of a scientific law will be analyzed in subchapter 11.1.
4 They are often called "phenomenological laws" in a philosophical tradition which never arrived at a conceptual distinction between appearances and manifestations.

5 The character of these singular events will be scrutinized more closely in subchapter 4.1.
6 That is why acceleration in (1.7) really comes out as $g^{(2)}$, in (1.8) but still only as g; that is why acceleration in (1.4) is, as a matter of fact, $g^{(2)}$, the scientific law involving nine, not eight, idealizations. My formulation (1.4) does not reveal the scientific law expressed by (1.6) as its background.

CHAPTER 4

1 I give up Popper's term "deducibility" from his definition *(N^o)* on the basis of my conclusion that the process of explanation of a universal law does not have the character of a deductive argument.

CHAPTER 5

1 The character of this necessity will be scrutinized in chapter eight.
2 This progression will be dealt with in chapter six.
3 I will deal with this in subchapter 5.3, and then in chapter seven.
4 I use the star to distinguish a natural kind's "cell–law" from the similar, yet somewhat different, law which was at work in the historically previous natural kind.

CHAPTER 6

1 I use here the notation from chapter three. Ellis uses the equation $f(t) = dI(t)/dt = dM(t)/dt$, where $f(t)$ stands for the instantaneous force, $I(t)$ is Newton's motive force impressed upon a body in time t, and $M(t)$ is the change of motion which the body undergoes in this time.
2 Hegel's *Wissenschaft der Logik* (1923) diverges in some places from the (1969) translation by A. V. Miller. In the quotations I, therefore, draw upon both of them.
3 I depart here from the tradition of symbolizing the dimension Q of a derived magnitude q by means of a multiplication of the dimensions $P, R, S, \ldots, Q = P{\cdot}R{\cdot}S \ldots$, of the fundamental (underived) magnitudes $p, r, s, \ldots$.
4 On Newton's attempts to determine the quality of forces see (McGuire 1968).
5 This, it seems, is implied by his view that only those properties are essential "which admit neither intensification nor remission of degrees, and which are found to belong to all bodies within the reach of our experiments" (1946, 398).
6 But, according to Hegel, even the second case still represents an inferior level of qualification of the quantitative as compared to Kepler's law $s^3 = at^2$, because here both sides of the relation are given in higher powers.
7 Hegel here draws upon the chemistry of his time; on this see, e.g., (Snelders 1993).
8 Ph. Mongin claims this in (1989).
9 "The process by which the values inside the money-system are determined by labour-time . . . drops *outside* circulation; it stays behind it as an acting ground and supposition" (1981a, 662; Marx and Engels 1975ff., Vol. 29, 176). H.-G. Backhaus stresses, correctly in my view, that "Value for Marx . . . is not an unmoving substance in un-differentiated rigidity (*Starrheit*), but a subject unfolding (*Entfaltendes*) itself in its differences" (1969, 146; 1980, 112).
10 Marx in (1983, 43) strictly distinguishes between the *quantity* (*Grösse*) and *quality* (substance) of the magnitude of value. Therefore, in my view, the standard English translation of the German "*Werthgrösse*" as "magnitude of value," e.g., in (Marx 1990, 88; 1967, Vol. 1, 102; Marx and Engels 1975ff., Vol. 35, 111) is not correct.
11 It represents a "negative" process interrupting and mediating the process of the production of value and surplus-value.
12 From the point of view of the dimensional analysis this implies that such a theory would be expressed by means of one dimension X.
13 I draw here upon Marx's ideas from Section 3 "Form of Value or Exchange-Value."

14 The sentential connective "or" stands here for "Either . . . or . . . , but only one of them."
15 On the coefficient appearing in the law of value see Quaas (1984).

CHAPTER 7

1 The category of substance-subject will be analyzed in chapter ten. A comparison of Marx's and Hegel's approach to it will be given in chapter eleven.
2 I do not analyze here the relation of Marx's *Capital* as a whole (first volume and the manuscript of the second and the third volume) to the manuscript *Theories of Surplus Value*.
3 The character of this universality will be analyzed in chapter ten.

CHAPTER 8

1 Objections to such a transfer were raised by K. S. Donnellan in (1983).
2 The same is claimed, e.g., by K. Fine in (1994); see also his (1995a) and (1995b).
3 This was pointed out also, e.g., by J. Hintikka in (1982).
4 An objection similar to this was raised in (Bigelow and Ellis and Lierse 1992).
5 I do not analyze the type of necessity expressed in scientific laws of the type L_h.
6 The concept of a natural kind I understand here as the sense of a natural kind-term.
7 From this and the previous note it follows that in my view names of natural kinds do have a (Fregean) sense.
8 Carnap's (1964) is a direct answer to A. Pap's (1964) and an indirect answer to his (1958) paper.
9 L stands here for an extensional language.
10 See on it Wilson (1979).

CHAPTER 9

1 One can reconstruct Mellor's approach to causal loops by using singular-event probabilities because he claims: "I need not identify the probability of effects with the frequency resulting from a statistical test of causation. Indeed, I cannot, since I do not take causation to relate only types or classes of events. On the contrary, to make sense of backward causation I have had to make causation to relate token events. So if causes have to increase the probability of their effects, they must be able to do it in a single case," (1981, 186) and because his approach to probability is based on the propensity interpretation of probability, where propensities are displayed in single trials (Mellor 1971, 66–67).
2 $N = 1000$ represents a relatively small number. The bacteria *Escherichia Coli* has a genome of several millions of "symbols," while the human genome contains nearly a thousand times more "symbols" (Eigen–Winkler-Oswatitsch 1992, 10).
3 Such an accidental process could not take place even outside the Earth in outer space (Eigen and Winkler-Oswatitsch 1992, 11).
4 As one of the *external* conditions of life on Earth Eigen mentions the carbon cycle which keeps the sun shining (Eigen and Schuster 1979, 2).
5 The words are "TAKE," "ADVANTAGE," "OF" and "MISTAKE"; the last being purportedly chosen so as to feed back to the first word.
6 As the next stage of the evolution of life Eigen views the becoming of a fully compartmentalized hypercycle (see Eigen and Gardiner and Schuster 1980; Eigen and Winkler-Oswatitsch 1992, 34–37). Because I approach the work of M. Eigen only from the point of view of the cyclic relation between cause and effect, I do not analyze the issue of compartmentalization.
7 For a critique see, e.g., (Riggs 1991).
8 This thought-experiment draws upon Rigg's thought-experiment mentioned in his (1991).
9 On this see, e.g., (Reck 1960; Rottenstreich 1974; Stern 1990).
10 On this see, e.g., (Eigen et al. 1991).

CHAPTER 10

1 Frege claims "In the case of thinking it is not really ideas that are connected, but things, properties, concepts, relations" (1906a, 174).
2 Frege used the term "realm of reference" ("*Reich der Bedeutung*") in (1919, 255).
3 Especially in chapter nineteen in (1981a) and in chapter three in (1981b).
4 "concept" is here understood as equivalent in meaning to Frege's "sense."
5 I. Hacking uses the term "lonely ego" to characterize such an approach to human beings (1975, 57). On this se also S. Ellenbogen's (1997).
6 In such a case it would not be *conscious reflection*, but only *reflection*, e.g., as in the case of the image of a ring embedded in wax.
7 On the emergence of the cognitive forms of thought in the ontogenesis of man see Piaget (1981); Piaget and Inhelder (1970). On their origin in the phylogenesis of man see Luria (1976), especially chapter one dealing with the sociohistorical formation of mind.
8 The general characteristics of logical categories are outlined by V. S. Stepin as follows (1985, 98):

> Any form of human cognition (every-day cognition, or science, or artistic appropriation of the world) in each epoch is realized in correspondence with the historically created categorial structures fixing the determinations of being, which were discovered by the previous development of cognition and practice.

These categories "determine the ultimately general scheme of understanding and transformation" of reality. They (1985, 101–108)

> appear as the basic structures of human consciousness. They are universal because any object can become the subject matter of action . . . Categories become, develop and function in the culture as a unified system of wholeness. Such a system appears as a generalized model of man's world, which is transmitted in the culture and is assimilated by individuals in the process of socialization. . . . By transmitting the accumulated social experience, passing it from generation to generation, the categories of culture ensure the reproduction of a certain kind of social life and of a certain type of human individual. They appear as, in a certain way, deep programs (structures) which predetermine the interlocking, reproduction and variation of the whole manifold of concrete forms and types of behavior and action, characterizing a certain type of social organization.

9 This contradicts Hegel's own claim in the *Encyclopaedia Logic*, when he wrote, correctly in my view, that the content of human thinking initially does not appear in the form of idea, but in the form of feeling, intuition, and representation (1840a, §2, 4; 1991, 25; 1959, 4).
10 The inclusion of *PS* into the scheme of the reproduction cycle of cognition enables us to understand anew the third book the "Notion," of Hegel's *Science of Logic*, where the "action of logic" changes into the "logic of action." Instead of Hegel's theory of the self-movement of notion it is necessary to provide a cluster of philosophical categories theoretically grasping the thought-forms by means of which human beings transform in mind reality in accordance with the aims, wants and values.
11 I do not reconstruct in this book that type of reproduction cycle which is "centered" around the empirical (phenomenological) type of scientific law.

CHAPTER 11

1 In the *Science of Logic* Hegel, because he treats problems of the organic in the cluster "Notion," provides a typology of scientific laws which is poorer than that found in the *Phenomenology*.
2 Hegel characterizes scientific laws in *Science of Logic* in a similar way. He states that "the realm of laws is the *quiet* (*ruhende*) image of the world of existence or manifestation" (1923, T. 2, 127; 1969, 503).
3 Nor is such a reconstruction provided in Part III of L. Nowak's (1974) treatment of the principles of the development of theoretical science, where he reconstructs several models of the development of a scientific law. In the last two of them he even considers the case when

the understanding of what the principal factors of a phenomenon are undergoes a profound change. But he claims that the universe of discourse does not change in the course of such a development of knowledge.

4 For a detailed analysis of Hegel's approach to the organic see O. Breidbach's (1982).

5 I draw here upon M. Wolff's (1995a; 1995b).

6 I draw here upon A. Nuzzo's excellent (1995) paper.

7 On Kant's views about ideas as schemes see his (B698; 1965, 55); for a commentary see H. Pilot's (1995) and T. Leiber's (1996).

8 For an overview of this criticism see H. Priest's (1987a).

9 This quotation appears in the second introduction to the *Encyclopaedia*, which is not translated in (Hegel 1959).

10 In §3 of the *Encyclopaedia Logic* Hegel does not view thoughts as a lower-level form of consciousness as compared to that of notion (1840a, 6; 1991, 26–27; 1959, 6–8).

11 On this merger see H. F. Fulda's (1990).

12 On this see N. Rottenstreich's (1974).

13 In the "Notion"-cluster, according to Kollman, "truer forms replace the category of causality, substance, mutual interaction," because they "have no validity (*Gelten*) any more" (1970, 413). Categories of this cluster, Hegel claims, replace "forms such as categories and determinations of reflection, whose finitude and untruth has demonstrated itself in logic" (1923, T. 2, 231; 1969, 592).

14 A detailed analysis of Marx's "Method of Political Economy" is given in Jánoska and Bondeli and Kindle and Hofer (1994).

15 For an analysis of Hegel's approach to contradiction by means of the symbols $+A$ and $-A$, see M. Wolff's (1981; 1986).

REFERENCES

(References in the text are to reprints and translations where these are given with the exception of Hegel's and Marx's works, where both the original and the translation is referred to. In the case of Marx as the primary source of reference is taken, wherever possible, the second edition of the MARX/ENGELS GESAMTAUSGABE (MEGA^2)).

Achinstein, P., and Barker, S. eds. 1969. *The Legacy of Logical Positivism.* Baltimore: Johns Hopkins Press.

Achinstein, P. 1981. *The Nature of Explanation.* Clarendon: Oxford.

Ackerman, J., and Wlotzka, B., and McCaskill, J.S. 1998. "In *vitro* DNA-based Predator-Prey System with Oscillatory Kinetics," *Bulletin of Mathematical Biology* 60: 329–354.

Aristotle 1941a. *Categories.* In McKeon 1941: 7–39.

—. 1941b. *Metaphysics.* In McKeon 1941: 686–934.

Armstrong, D.M. 1978. *Universals and Scientific Realism*, Vol. I, II. Cambridge: Cambridge University Press.

—. 1983. *What is a Law of Nature?* Cambridge: Cambridge University Press.

Backhaus, H.-G. 1969. "Zur Dialektik der Wertform," in Schmidt, A., ed., *Beiträge zur marxistischen Erkenntnistheorie.* Frankfurt a. Main, Suhrkamp: 128–151.

—. 1980. "On the Dialectics of the Value Form," in *Thesis Eleven*, Vol. 1. Cambridge, MA, M.I.T. Press: 99–120.

Backhaus, H.-G., and Reichelt, H. 1995. "Der politisch-ideologische Grundcharakter der Marx und Engels-Gesamtausgabe," in *MEGA-Studien*, 1994/2. Berlin, Dietz Verlag: 101–118.

Balashov, Y.V. 1992. "On the Evolution of Natural Laws," *The British Journal for the Philosophy of Science* 43: 343–370.

Beauchamp, T.L., ed. 1974. *Philosophical Problems of Causation.* Encino, CA: Dickenson.

Bigelow, J., and Ellis, B., and Lierse, C. 1992. "The World as One of a Kind," *The British Journal for the Philosophy of Science* 43: 371–388.

Breidbach, O. 1982. *Das Organische in Hegels Denken.* Würzburg: Königshausen und Neumann.

Bromberger, S. 1966. "Why Questions," in Colodny, R., ed., *Mind and Cosmos.* Pittsburgh, Pittsburgh University Press: 86–111.

Carnap, R. 1939. "Foundations of Logic and Mathematics," *International Encyclopedia of Unified Science*, Vol. I., No. 3. University of Chicago Press Chicago: viii–71. Reprinted in Neurath, O., and Carnap, R., and Morris, Ch., eds. 1965. *Foundations of the Unity of Science*, Vol. I (Volumes I–II of the Encyclopedia) Chicago, University of Chicago Press, Eleventh Impression: iii–71.

—. 1950. "Empiricism, Semantics, Ontology," *Revue International de Philosophie* 4: 20–40. Reprinted in Carnap 1956: 205–221.

—. 1956. *Meaning and Necessity.* Chicago: University of Chicago Press.

—. 1964. "Arthur Pap on Dispositions," in Schilpp 1964: 947–952,.

Çolodny, R.G. ed., 1965. *Beyond the Edge of Certainty.* Englewood Cliffs, N.J.: Prentice Hall.

Černík, V. 1977. Problém zákona v marxistickej metodológii vied (*The Problem of Scientific Law in the Marxist Methodology of Science*). Bratislava: Pravda Publishing House.

—. 1986. Systém kategórií materialistickej dialektiky (*The System of Categories of Materialistic Dialectics*). Bratislava: Pravda Publishing House.

Donnellan, K.S. 1983. "Kripke and Putnam on Natural Kind Terms," in Ginet, C. ed., *Knowledge and Mind.* Oxford, Oxford University Press: 84–104.

Doz, A. 1974. "Analyse de la marchandise chez Marx et théorie de la mesure chez Hegel," in D'Hondt, J., ed., *La logique de Marx.* Paris, PUF: 91–103.

Dretske, F. 1977. "Laws of Nature," *Philosophy of Science* 44: 248–268.

Dummett, M. 1964. "Bringing about the Past," in *Philosophical Review* 73: 338–359. Reprinted in Dummett 1978: 333–350.

—. 1978. *Truth and Other Enigmas*. Cambridge, MA: Harvard University Press.
—. 1981a. *Frege's Philosophy of Language*. 2nd edition. London: Duckworth.
—. 1981b. *The Interpretation of Frege's Philosophy*. London: Duckworth.
Eigen, M. 1971. "Selforganization of Matter and the Evolution of Biological Macromolecules," *Naturwissenschaften* 58: 465–523.
—. 1976. "Wie entsteht Information?" *Berichte der Bunsen Gesellschaft* 80: 1059–1081.
—. 1980. *Das Urgen*. Halle (Saale): Deutsche Akademie der Naturforscher Leopoldina.
—. 1983. "Leben aus der Retorte?" *Münchener Medizinische Wochenschrift* 125, Supplement 1: 125–135.
—. 1986. "The Physics of Evolution," *Chemical Scripta* 26: 13–26.
—. 1993a. "The Origin of Genetic Information," *Gene* 269: 42–49.
—. 1993b. "Virus Strains as Models of Molecular Evolution," *Medicinal Research Reviews* 13: 385–398.
—. 1994. "Selection and the Origin of Information," *International Review of Neurobiology* 37: 35–50.
Eigen, M. et al. 1981. "The Origin of Genetic Information," *Scientific American* 244: 88–118.
—. 1989. "How Old is the Genetic Code?," *Science* 244: 673–679.
—. 1991. "The Hypercycle. Coupling of RNA and Protein Biosynthesis in the Infection Cycle of an RNA Bacteriophage," *Biochemistry* 30: 11005–11018.
Eigen, M., and Gardiner, W.C. Jr., and Schuster, P. 1980. "Hypercycle and Compartments," *Journal of Theoretical Biology* 85: 407–411.
Eigen, M., and McCaskill, J., and Schuster, P. 1988. "Molecular Quasi-Species," *Journal of Physical Chemistry* 92: 6881–6891.
Eigen, M., and Schuster, P. 1979. *The Hypercycle*. Berlin: Springer Verlag.
—. 1982. "Stages of Emerging Life," *Journal of Molecular Evolution* 19: 47–61.
Eigen, M., and Winkler, R. 1974. "Ludus Vitalis," *Mannheimer Forum* 1973/74. Mannheim, Studienreihe Boehringer Mannheim: 53–139.
—. 1981a. *Laws of the Game*. New York: Knopf.
Eigen, M., and Winkler-Oswatitsch, R. 1981b. "Transfer-RNA: The Early Adaptor," *Naturwissenschaften* 68: 217–228.
—. 1981c. "Transfer-RNA, an Early Gene?" *Naturwissenschaften* 68: 282–292.
—. 1992. *Steps Towards Life*. Oxford: Oxford University Press.
Eigen, M. and Winkler-Oswatitsch, R., and Dress, A. 1988. "Statistical Geometry in Sequence Space," *Proceedings of the National Academy of Science USA* 85: 5913–5917.
Ellenbogen, S. 1997. "On the Link Between Frege's Platonic-Realistic Semantics and his Doctrine of Private Senses," *Philosophy* 72: 375–382.
Ellis, B. 1965. "The Origin and Nature of Newton's Laws of Motion," in Colodny 1965: 29–68.
Enç, B. 1975. "Necessary Properties and Linnaen Essentialism," *Canadian Journal of Philosophy* 5: 83–102.
—. 1986. "Essentialism without Individual Essences," in French and Uehling and Wettstein 1986: 403–425.
Feigl, H. 1969. "The Origin and Spirit of Logical Positivism," in Achinstein and Barker 1969: 3–24.
Feigl, H., and Scriven, M., and Maxwell, G., eds. 1958. *Minnesota Studies in the Philosophy of Science, Vol. II*. Minneapolis: University of Minnesota Press.
Fine, K. 1994. "Essence and Modality," in Tomberlin, J., ed., *Philosophical Perspectives 8*. Atascadero, Ridgeview: 1–16.
—. 1995a. "Senses of Essence," in Sinnott-Armstrong, W., ed., *Modality, Morality, and Belief*. Cambridge, Cambridge University Press: 53–73.
—. 1995b. "The Logic of Essence," *Journal of Philosophical Logic* 24: 241–273.
Fisk, M. 1970. "Are there Necessary Connections in Nature?" *Philosophy of Science* 37: 385–404.
Frege, G. 1885. "Review of H. Cohen, *Das Prinzip der Infentisimal-Methode und seine Geschichte*," *Zeitschrift für Philosophie und philosophische Kritik* 87: 324–329. Reprinted in Frege 1967: 99–102.

—. 1892. "Über Sinn und Bedeutung," *Zeitschrift für Philosophie und philosophische Kritik* 100: 25–50. Translated as "On Sense and Reference," in Geach and Black 1977: 56–78.
—. 1897. "Logik," Translated in Frege 1979: 126–151.
—. 1906a. "17 Kernsätze zur Logik," Translated in Frege 1979: 174–176.
—. 1906b. "Einleitung in die Logik," Translated in Frege 1979: 185–196.
—. 1918a. "Der Gedanke: eine logische Untersuchung," *Beiträge zur Philosophie des deutschen Idealismus* I: 58–77. Translated in Frege 1977: 1–30.
—. 1918b. "Die Verneinung: eine logische Untersuchung," in *Beiträge zur Philosophie des deutschen Idealismus* I: 143–157. Translated in Frege 1977: 31–53.
—. 1919. "Aufzeichnungen für Ludwig Darmstädter," Translated in Frege 1979: 253–257.
—. 1923. "Logische Allgemenheit," Translated in Frege 1979: 258–262.
—. 1924/25. "Erkenntnisquellen der Mathematik und der mathematischen Naturwissenschaften," Translated in Frege 1979: 267–274.
—. 1967. *Kleine Schriften*. Darmstadt: Wissenschaftliche Buchgesellschaft.
—. 1979. *Posthumous Writings*. Oxford: Basil Blackwell.
French, P.A., and Uehling, F.E., and Wettstein, H.K., eds. 1986. *Midwest Studies in Philosophy XI*. Minneapolis: University of Minnesota Press.
Fricke, Ch., and König, P., and Petersen, T., eds. 1995. *Das Recht der Vernunft*. Stuttgart: Frommann-Holzboog.
Fulda, H.F. 1990. "Spekulative Logik als die 'eigentliche Metaphysik'," in Pätzold, D., and Vanderjagt, A., eds. *Hegels Transformation der Metaphysik*. Köln, Dinter: 9–28.
Geach, P., and Black, M., eds. 1977. *Translations from the Philosophical Writings of Gottlob Frege*. Oxford: Basil Blackwell.
Glansdorff, P. 1977. "Origine et perspectives des structures dissipatives," in Haken, H., ed. *Synergetics* (Far From Equilibrium). Berlin, Springer Verlag: 1–11.
Gray, B.F., and Morley-Buchanan, T. 1985. "Some Criticism Concerning the Brusselator Model of an Oscillating Chemical Reaction," *Journal of Chemical Society* 81, F2, Part 1: 77–85.
Griese, A. 1977. "Philosophische Betrachtung über die Richtung der Zeit," in Wessel 1977: 114–131.
Hacking, I. 1975. *Why Does Language Matter to Philosophy?* Cambridge: Cambridge University Press.
Hanzel, I., and Cerník, V., and Viceník, J. 1994. "What is a Category?" *Metaphilosophy* 25: 181–193.
Hegel, G.W.F. 1840a. *Encyclopädie der philosophischen Wissenschaften im Grundrisse. Erster Theil. Die Logik*. In Hegel, G.W.F. *Werke, Bd. 6*. Henning, L. von, ed., Berlin: Duncker und Humblot.
—. 1840b. *Vorlesungen über die Geschichte der Philosophie. Erster Theil*. In Hegel, G.W.F. *Werke, Bd. 13*. Michelet, C.L., ed., Berlin: Duncker und Humblot.
—. 1842. *Vorlesungen über die Naturphilosophie als der Encyclopädie der philosophischen Wissenschaften im Grundrisse. Zweiter Theil*. In Hegel, G.W.F. *Werke, Bd. 7; erste Abtheilung*. Michelet, C.L., ed., Berlin: Duncker und Humblot.
—. 1913. *System der Sittlichkeit*. In Hegel, G.W.F. *Sämtliche Werke, Band VII*. Schriften zur Politik und Rechtsphilosophie. Lasson, G., ed., Leipzig, Felix Meiner: 417–503.
—. 1923. *Wissenschaft der Logik*, T. 1, 2. In Hegel, G.W.F. *Sämtliche Werke, Band IV., V.* Lasson, G., ed., Leipzig: Felix Meiner.
—. 1928. *Phänomenologie des Geistes*. In Hegel, G.W.F. *Sämtliche Werke, Band II*. Lasson, G., ed., Leipzig: Felix Meiner.
—. 1955. *Lectures on the History of Philosophy, Vol. I-III*. Translated by E.S. Halsdane and F.H. Simon. New York: Humanities Press.
—. 1959. *The Logic*. 2nd ed. Translated by W. Wallace. Oxford: Oxford University Press.
—. 1967. *The Phenomenology of Mind*. Translated by J.B. Bailie. New York: Harper and Row.
—. 1969. *The Science of Logic*. Translated by A.V. Miller. Atlantic Highlands: Humanities Press.
—. 1970a. *Philosophy of Nature*. Translated by A.V. Miller. Oxford: Clarendon.
—. 1970b. *Philosophy of Nature, Vol. 1, 2, 3*. Translated by M.J. Petry. London: Allen and Unwin.

—. 1970c. *Berliner Schriften 1818–1831.* Frankfurt a. Main: Suhrkamp.
—. 1979. *System of Ethical Life and First Philosophy of Spirit.* Translated by H.S. Harris and T.M. Knox. Albany: SUNY Press.
—. 1991. *Encyclopaedia Logic.* Translated by T.F. Geraets, and W.A. Suchting, and H.S. Harris. Indianopolis: Hackett.
Hempel, C.G. 1952. "Problems of Concept and Theory Formation in the Social Sciences," in *Science, Logic, and Human Rights*. Philadelphia, University of Pennsylvania Press: 65–86. Reprinted as "Typological Methods in Natural and Social Science," in Hempel 1965b: 155–172.
—. 1958. "Theoretician's Dilemma," in Feigl and Scriven and Maxwell 1958: 37–98. Reprinted in Hempel 1965b: 173–228.
—. 1964. "Postscript to the Studies in the Logic of Explanation," In Hempel 1965b: 291–296.
—. 1965a. "Aspects of Scientific Explanation," in Hempel 1965b: 331–496.
—. 1965b. *Aspects of Scientific Explanation.* New York: The Free Press.
—. 1966. *Philosophy of Natural Science.* Englewood Cliffs: Prentice Hall.
—. 1970. "On the Standard Conception of Scientific Theories," in *Minnesota Studies in the Philosophy of Science, Volume IV.* Radner, M., and Winokur, S., eds., Minneapolis, University of Minnesota Press: 142–163.
—. 1988. "Provisos," in Grünbaum, A., and Salmon, W.C., eds., *Limitations of Deductivism.* Berkeley, University of California Press: 19–36.
– ., and Oppenheim, P. 1948. "Studies in the Logic of Explanation," *Philosophy of Science* 15: 135–175. Reprinted in Hempel 1965b: 245–290.
Hintikka, J. 1982. "Is Alethic Modal Logic Possible?" *Acta Philosophica Fennica* 35: 89–105.
Ihmig, K.-N. 1993. "Hegel's Rejection of the Concept of Force," In Petry 1993: 399–414.
Ilyenkov, E.V. 1977. "Considerations on the Question of the Relation of Thinking and Language (Speech)," (in Russian) *Voprosy Filosofii* 30: 94–96.
Jammer, M. 1957. *Concepts of Force.* Cambridge, MA: Harvard University Press.
—. 1961. *Concepts of Mass.* Cambridge, MA: Harvard University Press.
Jánoska, J., and Bondeli, M., and Kindle, K., and Hoffer, M. 1994. *Das «Methodenkapitel» von Karl Marx.* Basel: Schwabe & Co.
Jeffrey, R.C. 1969. "Statistical Explanation vs. Statistical Inference," in Rescher, N., ed., *Essays in Honor of Carl G. Hempel.* Dordrecht, Reidel: 104–113.
Kant, I. 1965. *The Critique of Pure Reason.* Translated by N.K. Smith. New York: St Martin's Press.
Kneale, W. 1948. "What Can do Logic for Philosophy?" *Proceedings of the Aristotelian Society*, Supplementary Volume XXII: 155–166.
—. 1949 *Probability and Induction.* Oxford: Clarendon.
—. 1950 "Natural Laws and Contrary-to-Fact Conditionals," *Analysis* 10: 121–126. Reprinted in Beauchamp 1974: 46–49.
—. 1955. "A Reply to Mr. Bennett," *The British Journal for the Philosophy of Science* 8: 57–62.
—. 1961. "Universality and Necessity," *The British Journal for the Philosophy of Science* 12: 89–102. Reprinted in Beauchamp 1974: 53–62.
—. 1962. *The Development of Logic.* Oxford: Clarendon.
Kollman, Chr. E. 1970. "Über die Hegelsche Lehre," in Hegel 1970c: 390–415.
Kripke, S. 1963. "Semantical Considerations on Modal Logic," *Acta Philophica Fennica* 16: 83–94.
—. 1971. "Identity and Necessity," in Munitz, M., ed., *Identity and Individuation.* New York, New York University Press. Reprinted in Schwartz, S., ed., *Naming, Necessity and Natural Kinds.* Ithaca, Cornell University Press: 66–101.
—. 1980. *Naming and Necessity*. Oxford: Basil Blackwell.
Kyburg, H.E. Jr. 1965. "Comments," *Philosophy of Science* 32: 147–151.
Lefever, R., and Nicolis, G., and Borckmans, P. 1988. "The Brusselator: It does Oscillate All the Same," *Journal of Chemical Society* 84, F1, Part 4: 1013–1021.
Leiber, K. 1996. "Kategorien, Schemata und empirische Begriffe," *Kant Studien* 87: 1–41.
Lukács, G. 1976. *The Young Hegel.* Cambridge, MA: M.I.T. Press.
Luria, A.R. 1976. *Cognitive Development.* Cambridge, MA: Harvard University Press.

Mach, E. 1868. "Über die Definition der Masse," *Carls Repertorium der Experimentalphysik* 4: 355–359.
—. 1960. *The Science of Mechanics.* Sixth edition. La Salle: Open Court.
Mackie, J.L. 1980. *The Cement of the Universe.* Oxford: Clarendon.
Mandel, E. 1985. "Partially Independent Variables and the Internal Logic in Classical Marxist Economic Analysis," *Social Science Information* 4: 485–505.
Marx, K. 1967. *Capital, Vol. 1–3.* New York: International Publishers.
—. 1969. "Randglosen zur Adolph Wagners 'Lehrbuch der politischen Ökonomie'," *Marx and Engels Werke, Bd. 19*: 355–383. Berlin: Dietz Verlag. Translated in Marx and Engels 1975ff., Vol. 24: 531–559.
—. 1970. "Marx an Engels. 2. April 1858," *Marx and Engels Werke, Bd. 29*: 309–318. Berlin: Dietz Verlag. Translated in Marx and Engels 1975ff., Vol. 40: 296–304.
—. 1976. *Ökonomische Manuskripte 1857/58* (Grundrisse), Teil 1. MEGA^2 II/1.1. Berlin: Dietz. Translated in Marx and Engels 1975ff., Vol. 28.
—. 1977. "[Marx's Précis of Hegel's Doctrine of Being in the Minor Logic]," *International Review of Modern History* 22: 428–431.
—. 1979. *Zur Kritik der politischen Ökonomie* (Manuskript 1861–1863), Teil 4. MEGA^2 II/3.4. Berlin: Dietz Verlag. Translated in Marx and Engels 1975ff., Vol. 32.
—. 1980a. *Zur Kritik der politischen Ökonomie. Urtext.* In Marx 1980c: 19–94. Translated in Marx and Engels 1975ff., Vol. 29: 430–507.
—. 1980b. *Zur Kritik der politischen Ökonomie*, Erstes Heft. Berlin 1859: Franz Duncker. In Marx 1980c: 99–255. Translated in Marx and Engels 1975ff., Vol. 29: 257–417.
—. 1980c. *Ökonomische Manuskripte und Schriften 1858–1861.* MEGA^2 II/2. Berlin: Dietz Verlag.
—. 1980d. *Zur Kritik der politischen Ökonomie* (Manuskript 1861–1863), Teil 5. MEGA^2 II/3.5. Berlin: Dietz Verlag. Translated in Marx and Engels 1975ff., Vol. 33.
—. 1981a. *Ökonomische Manuskripte 1857/58* (Grundrisse), Teil 2. MEGA^2 II/1.2. Berlin: Dietz Verlag. Translated in Marx and Engels 1975ff., Vol. 28, 29.
—. 1981b. "The Circulation Process of Capital (Manuscript II)," (In Russian) in Marx, K., and Engels, F. *Works, Vol. 50.* Moscow, Izdatelstvo polititcheskoj literatury: 5–302.
—. 1982a. *Zur Kritik der politischen Ökonomie* (Manuskript 1861–63), Teil 6. MEGA^2 II/3.6. Berlin: Dietz Verlag. Translated in Marx and Engels 1975ff., Vol. 34.
—. 1982b. *Zur Kritik der Hegelschen Rechtsphilosophie.* MEGA^2 I/2 (Werke, Artikel, Entwürfe März 1843–August 1844). Berlin, Dietz Verlag: 3–137. Translated in Marx and Engels 1975ff., Vol. 3: 3–130.
—. 1983. *Das Kapital.* Erster Band. Hamburg 1867: Otto Meissner. MEGA^2 II/5. Berlin: Dietz Verlag.
—. 1984. "The Future Results of British Rule in India," MEGA^2 I/12 (Werke, Artikel, Entwürfe, Januar bis Dezember 1853). Berlin, Dietz Verlag: 248–253. Translated in Marx and Engels 1975ff., Vol. 12: 217–222.
—. 1987a. "Ergänzungen und Veränderungen zum ersten Band des 'Kapitals' (Manuskript December 1871–Januar 1872)," MEGA^2 II/6. Berlin, Dietz Verlag: 1–54.
—. 1987b. *Das Kapital.* Erster Band. Zweite verbesserte Auflage. Hamburg 1872: Otto Meissner. MEGA^2 II/6. Berlin: Dietz Verlag.
—. 1988a. "Das Productionsprocess des Capitals [Einzelne Seiten]," in Marx 1988d: 6–23. Translated in Marx and Engels 1975ff., Vol. 34, 339–354.
—. 1988b. "Sechstes Kapitel. Resultate des unmittelbaren Produktionsprozesses," in Marx 1988d: 24–135. Translated in Marx and Engels 1975ff., Vol. 34: 355–466.
—. 1988c. "Der Zirkulationsprozess des Kapitals. (Manuskript I)," in Marx 1988d: 137–381.
—. 1988d. *Ökonomische Manuskripte 1863–1867,* Teil 1. MEGA^2 II/4.1. Berlin: Dietz Verlag.
—. 1989. *Le Capital.* Paris 1872–1875: Maurice Lechâtre. MEGA^2 II/7. Berlin: Dietz Verlag.
—. 1990. *Capital.* Volume I. Translated from the third German Edition by Samuel Moore and Edward Aveling. London 1887: Swan Sonnenschein, Lowrey & Co. MEGA^2 II/9. Berlin: Dietz Verlag.

—. 1992. *Das Kapital (Ökonomisches Manuskript 1863–1865). Drittes Buch.* In *Ökonomische Manuskripte 1863–1867,* Teil 2. MEGA² II/4.2. Berlin-Amsterdam: Dietz Verlag-Internationales Institut für Sozialgeschichte.

Marx, K., and Engels, F. 1975ff. *Collected Works.* New York: International Publisher.

McGuire, J.E. 1968. "Force, Active Principles and Newton's Invisible Realm," *Ambix* 15: 154–208. Reprinted in McGuire 1996: 190–237.

—. 1996. *Tradition and Innovation.* Dordrecht: Kluwer.

McKeon, R., ed. 1941. *The Basic Works of Aristotle.* New York: Random House.

Mellor, D.H. 1971. *Matter of Chance.* Cambridge: Cambridge University Press.

—. 1980. "Necessities and Universals in Natural Laws," in Mellor, D.H., ed., *Science, Belief and Behaviour.* Cambridge, Cambridge University Press: 105–125.

—. 1981. *Real Time.* Cambridge: Cambridge University Press.

Metzler, H. 1977. "Kategorie Maß – Mathematik und Messen," *Deutsche Zeitschrift für Philosophie* 25: 596–600.

Mill, J.S. 1950. *A System of Logic.* Reprinted in Nagel, E., ed., *John Stuart Mill's Philosophy of Scientific Method.* New York: Hafner.

Mongin, Ph. 1989. "Les deux théories marxiennes de la valeur-travail et le problème de la mesure immanente,"*Archives de Philosophie* 52: 247–266.

Murray, P. 1990. *Marx's Theory of Scientific Knowledge.* Atlantic Highlands: Humanities Press.

Nerlich, G.C., and Suchting, W.A. 1967. "Popper on Law and Natural Necessity," *The British Journal for the Philosophy of Science* 18: 233–266.

Newton, I. 1687. *Philosophiae Naturalis Principia Mathematica.* Londini: Josephi Streater. (Reproduction, London around 1960: Dawson).

—. 1946. *Sir Isaac Newton's Mathematical Principles of Natural Philosophy and his System of the World.* (Translation A. Motte, revised by F. Cajori). Berkeley: University of California Press.

Nicolis, G., and Prigogine, I. 1977. *Self-Organization in Nonequilibrium Systems.* New York: Wiley Interscience.

—. 1989. *Exploring Complexity.* New York: Freeman.

Nowak, I. 1974. "Koncepcja dialektycznej korespondencji (The Conception of Dialectical Correspondence)," in Kmita, J., ed., *Metodologizcne implikacje epistemologii marksistowskiej* (The Methodological Implications of Marxist Epistemology). Warszawa, PWN: 252–256.

Nowak, I., and Nowak, L. 1985. "The Position of the Theory of Mind in the Categorial Interpretation of Mind," in Chisholm, R. et al., *Philosophy of Mind.* Wien, Halder: 264–270.

Nowak, L. 1971. *U podstaw Marksowskiej metodologii nauk* (Foundations of Marxian Methodology of Science). Warszawa: PWN.

—. 1972. "Laws of Science, Theory, Measurement," *Philosophy of Science* 39: 192–201.

—. 1974. *Zasady marksistowskiej filozofii nauki* (The Principles of the Marxist Philosophy of Science). Warszawa: PWN.

—. 1977. *U podstaw dialektyki Marksowskiej* (At the Basis of Marxian Dialectics). Warszawa: PWN.

—. 1980. *The Structure of Idealization.* Dordrecht: Reidel.

Nuzzo, A. 1995. "'Idee' bei Kant und Hegel," in Fricke and König and Petersen 1995: 81–120.

Otwinowski, M. 1990. "Structural Instability and Sustained Oscillations in an Extended Brusselator," *Canadian Journal of Physics* 68: 743–750.

Pap, A. 1958. "Disposition Concepts and Extensional Logic," In Feigl and Scriven and Maxwell 1958: 196–224.

—. 1964. "Reduction Sentences and Disposition Concepts," In Schilpp 1964: 559–598.

Pechmann, A. von 1980. *Die Kategorie des Maßes in Hegels »Wissenschaft der Logik«.* Köln: Pahl Rugenstein.

Petry, M.J., ed. 1993. *Hegel and Newtonianism.* Dordrecht: Kluwer.

Piaget, J. 1981. *The Psychology of Intelligence.* Tottowa: Littlefield.

Piaget, J., and Inhelder, B. 1970. *The Early Growth of Logic in the Child.* New York: Humanities Press.

Pilot, H. 1995. "Die Vernunftideen als Analoga von Schemata der Sinnlichkeit," in Fricke and König and Petersen 1995: 155–192.

Plato 1961. "The Sophist," in *Parmenides, Theaitetos, Sophist, Statesman.* New York, E.P. Dutton: 157–225.

Poincaré, H. 1911. "L'Evolution des Lois," *Scientia* 9: 275–292.

Popper, K.R. 1948. "What Can do Logic for Philosophy?" *Proceedings of the Aristotelian Society*, Supplementary Volume XXII: 141–154.

—. 1949. "A Note on Natural Laws and So-called 'Contrary-to-Fact Conditionals'," *Mind* LVIII: 62–66. Reprinted in Beauchamp 1974: 42–45.

—. 1967. "A Revised Definition of Natural Necessity," *The British Journal for the Philosophy of Science* 18: 316–321. Reprinted in Beauchamp 1974: 66–71.

—. 1968. *The Logic of Scientific Discovery*. London: Hutchinson.

Priest, S. 1987a. "Introduction," in Priest 1987b: 1–48.

Priest, S., ed. 1987b. *Hegel's Critique of Kant.* Oxford: Clarendon.

Prigogine, I. 1969. "Structure, Dissipation and Life," in Marois, M., ed., *Theoretical Physics and Biology*. Amsterdam, North Holland: 23–52.

—. 1980. *From Being to Becoming*. San Francisco: Freeman.

—. 1996. *La Fin des Certitudes.* Paris: O. Jacob.

Prigogine, I., and Stengers, I. 1984. *Order out of Chaos*. New York: Random House.

Quaas, G. 1984. "Zum Verhältniss von Wert und Preis aus mathematischer Sicht," *Wirtschaftswissenschaft* 32: 1649–1658.

—. 1988. "Wertgrössenbestimmung unter extremen Produktionsbedingungen," *Wirtschaftswissenschaft* 36: 1311–1322.

Reck, A.J. 1960. "Substance, Subject and Dialectic," in *Tulane Studies in Philosophy IX*. New Orleans, Tulane University: 109–133.

Reichenbach, H. 1928. *Philosophie der Raum-Zeit Lehre.* Berlin: W. de Gruyter und Co.

—. 1957. *The Philosophy of Space and Time.* New York: Dover.

—. 1958. *The Direction of Time.* Berkeley: University of California Press.

Renn, J. et al. 1998. *Hunting the White Elephant.* When and how did Galileo discover the law of fall? Preprint 97 of the Max Planck Institute of the History of Science: Berlin.

Riggs, P.J. 1991. "A Critique of Mellor's Argument against Backward Causation," *The British Journal for the Philosophy of Science* 42: 75–86.

Rottenstreich, N. 1974: *From Substance to Subject.* The Hague: Nijhoff.

Ruben, D.-H. 1979. "Marxism and Dialectics," in Mepham, J., and Ruben, D.-H., eds., *Issues in Marxist Philosophy, Vol. I.* Brighton, Harvester: 37–86.

Salmon, W.C. 1990. *Four Decades of Scientific Explanation*. Minneapolis: University of Minnesota Press.

Sarlemijn, A. 1993. "Pendulums in Newtonian Mechanics," In Petry 1993: 267–289.

Schilpp, P.A., ed. 1964. *The Philosophy of Rudolf Carnap.* La Salle: Open Court.

Schlesinger, G.N. 1987. "Nomic Necessity and Contingency," *The Philosophical Quarterly* 37: 379–391.

Schlick, M. 1979a. "Form and Content," in Schlick 1979b: 285–369.

—. 1979b. *Philosophical Papers, Vol. II.* Dordrecht: Reidel.

Schuster, P. 1981. "Prebiotic Evolution," in Gutfreund, H., ed., *Biochemical Evolution.* London, Cambridge University Press: 15–87.

Scriven, M. 1956. "Randomness and Causal Order," *Analysis* XVII: 5–9.

Snelders, H.A.M. 1993. "The Significance of Hegel's Treatment of Chemical Affinity," In Petry 1993: 631–643.

Stepin, V.S. 1985. "Socio-Cultural Dependence of the Heuristic Functions of Philosophy in Scientific Knowledge," (In Russian) in Stepin, V.S., ed., *The Leninist Theory of Reflection as the Methodology of Scientific Knowledge*. Minsk, University Publishing House: 88–124.

Stern, R. 1990. *Hegel, Kant and the Structure of the Object.* London: Routledge.

Suchting, W.A. 1969. "Popper's Revised Definition of Natural Necessity," *The British Journal for the Philosophy of Science* 20: 349–356.

Swoyer, C. 1982. "The Nature of Natural Laws," *Australasian Journal of Philosophy* 60: 203–223.
Tooley, M. 1977. "The Nature of Laws," *Canadian Journal of Philosophy* 7: 667–698.
Treder, H.-J. 1977. "Newtons Zeittheorie und die Richtung der Zeit," in Wessel 1977: 109–113.
Wahsner, R. 1977. "Struktur und Bewegung – Erkenntnis und Wirklichkeit," in Wessel 1977: 132–150.
Wessel, K.-F., ed. 1977. *Struktur und Prozess*. Berlin: Deutscher Verlag der Wissenschaft.
Will, U. 1984. "Nomische Notwendigkeit," *Zeitschrift für philosophische Forschung* 38: 582–592.
Wilson, M. 1979. "Maxwell Condition-Goodman Problem," *The British Journal for the Philosophy of Science* 30: 107–123.
Winkler-Oswatitsch, R., and Dress, A., and Eigen, M. 1986. "Comparative Sequence Analysis," *Chemica Scripta* 26B: 59–66.
Wolff, M. 1981. *Der Begriff des Widerspruchs*. Königstein/Ts.: Hain.
—. 1986. "Über Hegels Lehre vom Widerspruch," in Henrich, D., ed., *Hegels Wissenschaft der Logik*. Stuttgart, Klett-Cotta: 107–128.
—. 1995a. "Was ist formale Logik?" in Fricke and König and Petersen 1995: 19–31.
—. 1995b. *Die Vollständigkeit der kantischen Urteilstafel*. Frankfurt a. Main: Vittorio Klostermann.

INDEX

Boston Studies in the Philosophy of Science

Editor: Robert S. Cohen, *Boston University*

1. M.W. Wartofsky (ed.): *Proceedings of the Boston Colloquium for the Philosophy of Science, 1961/1962.* [Synthese Library 6] 1963 ISBN 90-277-0021-4
2. R.S. Cohen and M.W. Wartofsky (eds.): *Proceedings of the Boston Colloquium for the Philosophy of Science, 1962/1964.* In Honor of P. Frank. [Synthese Library 10] 1965 ISBN 90-277-9004-0
3. R.S. Cohen and M.W. Wartofsky (eds.): *Proceedings of the Boston Colloquium for the Philosophy of Science, 1964/1966.* In Memory of Norwood Russell Hanson. [Synthese Library 14] 1967 ISBN 90-277-0013-3
4. R.S. Cohen and M.W. Wartofsky (eds.): *Proceedings of the Boston Colloquium for the Philosophy of Science, 1966/1968.* [Synthese Library 18] 1969 ISBN 90-277-0014-1
5. R.S. Cohen and M.W. Wartofsky (eds.): *Proceedings of the Boston Colloquium for the Philosophy of Science, 1966/1968.* [Synthese Library 19] 1969 ISBN 90-277-0015-X
6. R.S. Cohen and R.J. Seeger (eds.): *Ernst Mach, Physicist and Philosopher.* [Synthese Library 27] 1970 ISBN 90-277-0016-8
7. M. Čapek: *Bergson and Modern Physics.* A Reinterpretation and Re-evaluation. [Synthese Library 37] 1971 ISBN 90-277-0186-5
8. R.C. Buck and R.S. Cohen (eds.): *PSA 1970.* Proceedings of the 2nd Biennial Meeting of the Philosophy and Science Association (Boston, Fall 1970). In Memory of Rudolf Carnap. [Synthese Library 39] 1971 ISBN 90-277-0187-3; Pb 90-277-0309-4
9. A.A. Zinov'ev: *Foundations of the Logical Theory of Scientific Knowledge (Complex Logic).* Translated from Russian. Revised and enlarged English Edition, with an Appendix by G.A. Smirnov, E.A. Sidorenko, A.M. Fedina and L.A. Bobrova. [Synthese Library 46] 1973 ISBN 90-277-0193-8; Pb 90-277-0324-8
10. L. Tondl: *Scientific Procedures.* A Contribution Concerning the Methodological Problems of Scientific Concepts and Scientific Explanation.Translated from Czech. [Synthese Library 47] 1973 ISBN 90-277-0147-4; Pb 90-277-0323-X
11. R.J. Seeger and R.S. Cohen (eds.): *Philosophical Foundations of Science.* Proceedings of Section L, 1969, American Association for the Advancement of Science. [Synthese Library 58] 1974 ISBN 90-277-0390-6; Pb 90-277-0376-0
12. A. Gr~nbaum: *Philosophical Problems of Space and Times.* 2nd enlarged ed. [Synthese Library 55] 1973 ISBN 90-277-0357-4; Pb 90-277-0358-2
13. R.S. Cohen and M.W. Wartofsky (eds.): *Logical and Epistemological Studies in Contemporary Physics.* Proceedings of the Boston Colloquium for the Philosophy of Science, 1969/72, Part I. [Synthese Library 59] 1974 ISBN 90-277-0391-4; Pb 90-277-0377-9
14. R.S. Cohen and M.W. Wartofsky (eds.): *Methodological and Historical Essays in the Natural and Social Sciences.* Proceedings of the Boston Colloquium for the Philosophy of Science, 1969/72, Part II. [Synthese Library 60] 1974 ISBN 90-277-0392-2; Pb 90-277-0378-7
15. R.S. Cohen, J.J. Stachel and M.W. Wartofsky (eds.): *For Dirk Struik.* Scientific, Historical and Political Essays in Honor of Dirk J. Struik. [Synthese Library 61] 1974 ISBN 90-277-0393-0; Pb 90-277-0379-5
16. N. Geschwind: *Selected Papers on Language and the Brains.* [Synthese Library 68] 1974 ISBN 90-277-0262-4; Pb 90-277-0263-2
17. B.G. Kuznetsov: *Reason and Being.* Translated from Russian. Edited by C.R. Fawcett and R.S. Cohen. 1987 ISBN 90-277-2181-5

Boston Studies in the Philosophy of Science

18. P. Mittelstaedt: *Philosophical Problems of Modern Physics.* Translated from the revised 4th German edition by W. Riemer and edited by R.S. Cohen. [Synthese Library 95] 1976
ISBN 90-277-0285-3; Pb 90-277-0506-2
19. H. Mehlberg: *Time, Causality, and the Quantum Theory.* Studies in the Philosophy of Science. Vol. I: *Essay on the Causal Theory of Time.* Vol. II: *Time in a Quantized Universe.* Translated from French. Edited by R.S. Cohen. 1980 Vol. I: ISBN 90-277-0721-9; Pb 90-277-1074-0
Vol. II: ISBN 90-277-1075-9; Pb 90-277-1076-7
20. K.F. Schaffner and R.S. Cohen (eds.): *PSA 1972.* Proceedings of the 3rd Biennial Meeting of the Philosophy of Science Association (Lansing, Michigan, Fall 1972). [Synthese Library 64] 1974 ISBN 90-277-0408-2; Pb 90-277-0409-0
21. R.S. Cohen and J.J. Stachel (eds.): *Selected Papers of LÇon Rosenfeld.* [Synthese Library 100] 1979 ISBN 90-277-0651-4; Pb 90-277-0652-2
22. M. Čapek (ed.): *The Concepts of Space and Time.* Their Structure and Their Development. [Synthese Library 74] 1976 ISBN 90-277-0355-8; Pb 90-277-0375-2
23. M. Grene: *The Understanding of Nature.* Essays in the Philosophy of Biology. [Synthese Library 66] 1974 ISBN 90-277-0462-7; Pb 90-277-0463-5
24. D. Ihde: *Technics and Praxis.* A Philosophy of Technology. [Synthese Library 130] 1979
ISBN 90-277-0953-X; Pb 90-277-0954-8
25. J. Hintikka and U. Remes: *The Method of Analysis.* Its Geometrical Origin and Its General Significance. [Synthese Library 75] 1974 ISBN 90-277-0532-1; Pb 90-277-0543-7
26. J.E. Murdoch and E.D. Sylla (eds.): *The Cultural Context of Medieval Learning.* Proceedings of the First International Colloquium on Philosophy, Science, and Theology in the Middle Ages, 1973. [Synthese Library 76] 1975 ISBN 90-277-0560-7; Pb 90-277-0587-9
27. M. Grene and E. Mendelsohn (eds.): *Topics in the Philosophy of Biology.* [Synthese Library 84] 1976 ISBN 90-277-0595-X; Pb 90-277-0596-8
28. J. Agassi: *Science in Flux.* [Synthese Library 80] 1975
ISBN 90-277-0584-4; Pb 90-277-0612-3
29. J.J. Wiatr (ed.): *Polish Essays in the Methodology of the Social Sciences.* [Synthese Library 131] 1979 ISBN 90-277-0723-5; Pb 90-277-0956-4
30. P. Janich: *Protophysics of Time.* Constructive Foundation and History of Time Measurement. Translated from German. 1985 ISBN 90-277-0724-3
31. R.S. Cohen and M.W. Wartofsky (eds.): *Language, Logic, and Method.* 1983
ISBN 90-277-0725-1
32. R.S. Cohen, C.A. Hooker, A.C. Michalos and J.W. van Evra (eds.): *PSA 1974.* Proceedings of the 4th Biennial Meeting of the Philosophy of Science Association. [Synthese Library 101] 1976 ISBN 90-277-0647-6; Pb 90-277-0648-4
33. G. Holton and W.A. Blanpied (eds.): *Science and Its Public.* The Changing Relationship. [Synthese Library 96] 1976 ISBN 90-277-0657-3; Pb 90-277-0658-1
34. M.D. Grmek, R.S. Cohen and G. Cimino (eds.): *On Scientific Discovery.* The 1977 Erice Lectures. 1981 ISBN 90-277-1122-4; Pb 90-277-1123-2
35. S. Amsterdamski: *Between Experience and Metaphysics.* Philosophical Problems of the Evolution of Science. Translated from Polish. [Synthese Library 77] 1975
ISBN 90-277-0568-2; Pb 90-277-0580-1
36. M. Marković and G. Petrović (eds.): *Praxis.* Yugoslav Essays in the Philosophy and Methodology of the Social Sciences. [Synthese Library 134] 1979
ISBN 90-277-0727-8; Pb 90-277-0968-8

Boston Studies in the Philosophy of Science

37. H. von Helmholtz: *Epistemological Writings.* The Paul Hertz / Moritz Schlick Centenary Edition of 1921. Translated from German by M.F. Lowe. Edited with an Introduction and Bibliography by R.S. Cohen and Y. Elkana. [Synthese Library 79] 1977
ISBN 90-277-0290-X; Pb 90-277-0582-8
38. R.M. Martin: *Pragmatics, Truth and Language.* 1979
ISBN 90-277-0992-0; Pb 90-277-0993-9
39. R.S. Cohen, P.K. Feyerabend and M.W. Wartofsky (eds.): *Essays in Memory of Imre Lakatos.* [Synthese Library 99] 1976 ISBN 90-277-0654-9; Pb 90-277-0655-7
40. Not published.
41. Not published.
42. H.R. Maturana and F.J. Varela: *Autopoiesis and Cognition.* The Realization of the Living. With a Preface to "Autopoiesis' by S. Beer. 1980 ISBN 90-277-1015-5; Pb 90-277-1016-3
43. A. Kasher (ed.): *Language in Focus: Foundations, Methods and Systems.* Essays in Memory of Yehoshua Bar-Hillel. [Synthese Library 89] 1976
ISBN 90-277-0644-1; Pb 90-277-0645-X
44. T.D. Thao: *Investigations into the Origin of Language and Consciousness.* 1984
ISBN 90-277-0827-4
45. F.G.-I. Nagasaka (ed.): *Japanese Studies in the Philosophy of Science.* 1997
ISBN 0-7923-4781-1
46. P.L. Kapitza: *Experiment, Theory, Practice.* Articles and Addresses. Edited by R.S. Cohen. 1980 ISBN 90-277-1061-9; Pb 90-277-1062-7
47. M.L. Dalla Chiara (ed.): *Italian Studies in the Philosophy of Science.* 1981
ISBN 90-277-0735-9; Pb 90-277-1073-2
48. M.W. Wartofsky: *Models.* Representation and the Scientific Understanding. [Synthese Library 129] 1979 ISBN 90-277-0736-7; Pb 90-277-0947-5
49. T.D. Thao: *Phenomenology and Dialectical Materialism.* Edited by R.S. Cohen. 1986
ISBN 90-277-0737-5
50. Y. Fried and J. Agassi: *Paranoia.* A Study in Diagnosis. [Synthese Library 102] 1976
ISBN 90-277-0704-9; Pb 90-277-0705-7
51. K.H. Wolff: *Surrender and Cath.* Experience and Inquiry Today. [Synthese Library 105] 1976
ISBN 90-277-0758-8; Pb 90-277-0765-0
52. K. Kosík: *Dialectics of the Concrete.* A Study on Problems of Man and World. 1976
ISBN 90-277-0761-8; Pb 90-277-0764-2
53. N. Goodman: *The Structure of Appearance.* [Synthese Library 107] 1977
ISBN 90-277-0773-1; Pb 90-277-0774-X
54. H.A. Simon: *Models of Discovery* and Other Topics in the Methods of Science. [Synthese Library 114] 1977 ISBN 90-277-0812-6; Pb 90-277-0858-4
55. M. Lazerowitz: *The Language of Philosophy.* Freud and Wittgenstein. [Synthese Library 117] 1977 ISBN 90-277-0826-6; Pb 90-277-0862-2
56. T. Nickles (ed.): *Scientific Discovery, Logic, and Rationality.* 1980
ISBN 90-277-1069-4; Pb 90-277-1070-8
57. J. Margolis: *Persons and Mind.* The Prospects of Nonreductive Materialism. [Synthese Library 121] 1978 ISBN 90-277-0854-1; Pb 90-277-0863-0
58. G. Radnitzky and G. Andersson (eds.): *Progress and Rationality in Science.* [Synthese Library 125] 1978 ISBN 90-277-0921-1; Pb 90-277-0922-X
59. G. Radnitzky and G. Andersson (eds.): *The Structure and Development of Science.* [Synthese Library 136] 1979 ISBN 90-277-0994-7; Pb 90-277-0995-5

Boston Studies in the Philosophy of Science

60. T. Nickles (ed.): *Scientific Discovery.* Case Studies. 1980 ISBN 90-277-1092-9; Pb 90-277-1093-7
61. M.A. Finocchiaro: *Galileo and the Art of Reasoning.* Rhetorical Foundation of Logic and Scientific Method. 1980 ISBN 90-277-1094-5; Pb 90-277-1095-3
62. W.A. Wallace: *Prelude to Galileo.* Essays on Medieval and 16th-Century Sources of Galileo's Thought. 1981 ISBN 90-277-1215-8; Pb 90-277-1216-6
63. F. Rapp: *Analytical Philosophy of Technology.* Translated from German. 1981 ISBN 90-277-1221-2; Pb 90-277-1222-0
64. R.S. Cohen and M.W. Wartofsky (eds.): *Hegel and the Sciences.* 1984 ISBN 90-277-0726-X
65. J. Agassi: *Science and Society.* Studies in the Sociology of Science. 1981 ISBN 90-277-1244-1; Pb 90-277-1245-X
66. L. Tondl: *Problems of Semantics.* A Contribution to the Analysis of the Language of Science. Translated from Czech. 1981 ISBN 90-277-0148-2; Pb 90-277-0316-7
67. J. Agassi and R.S. Cohen (eds.): *Scientific Philosophy Today.* Essays in Honor of Mario Bunge. 1982 ISBN 90-277-1262-X; Pb 90-277-1263-8
68. W. Krajewski (ed.): *Polish Essays in the Philosophy of the Natural Sciences.* Translated from Polish and edited by R.S. Cohen and C.R. Fawcett. 1982 ISBN 90-277-1286-7; Pb 90-277-1287-5
69. J.H. Fetzer: *Scientific Knowledge.* Causation, Explanation and Corroboration. 1981 ISBN 90-277-1335-9; Pb 90-277-1336-7
70. S. Grossberg: *Studies of Mind and Brain.* Neural Principles of Learning, Perception, Development, Cognition, and Motor Control. 1982 ISBN 90-277-1359-6; Pb 90-277-1360-X
71. R.S. Cohen and M.W. Wartofsky (eds.): *Epistemology, Methodology, and the Social Sciences.* 1983. ISBN 90-277-1454-1
72. K. Berka: *Measurement.* Its Concepts, Theories and Problems. Translated from Czech. 1983 ISBN 90-277-1416-9
73. G.L. Pandit: *The Structure and Growth of Scientific Knowledge.* A Study in the Methodology of Epistemic Appraisal. 1983 ISBN 90-277-1434-7
74. A.A. Zinov'ev: *Logical Physics.* Translated from Russian. Edited by R.S. Cohen. 1983 [*see also* Volume 9] ISBN 90-277-0734-0
75. G-G. Granger: *Formal Thought and the Sciences of Man.* Translated from French. With and Introduction by A. Rosenberg. 1983 ISBN 90-277-1524-6
76. R.S. Cohen and L. Laudan (eds.): *Physics, Philosophy and Psychoanalysis.* Essays in Honor of Adolf Gr~nbaum. 1983 ISBN 90-277-1533-5
77. G. Bhme, W. van den Daele, R. Hohlfeld, W. Krohn and W. SchÎfer: *Finalization in Science.* The Social Orientation of Scientific Progress. Translated from German. Edited by W. SchÎfer. 1983 ISBN 90-277-1549-1
78. D. Shapere: *Reason and the Search for Knowledge.* Investigations in the Philosophy of Science. 1984 ISBN 90-277-1551-3; Pb 90-277-1641-2
79. G. Andersson (ed.): *Rationality in Science and Politics.* Translated from German. 1984 ISBN 90-277-1575-0; Pb 90-277-1953-5
80. P.T. Durbin and F. Rapp (eds.): *Philosophy and Technology.* [*Also* Philosophy and Technology Series, Vol. 1] 1983 ISBN 90-277-1576-9
81. M. Marković: *Dialectical Theory of Meaning.* Translated from Serbo-Croat. 1984 ISBN 90-277-1596-3
82. R.S. Cohen and M.W. Wartofsky (eds.): *Physical Sciences and History of Physics.* 1984. ISBN 90-277-1615-3

Boston Studies in the Philosophy of Science

83. É. Meyerson: *The Relativistic Deduction.* Epistemological Implications of the Theory of Relativity. Translated from French. With a Review by Albert Einstein and an Introduction by Milić Čapek. 1985 ISBN 90-277-1699-4
84. R.S. Cohen and M.W. Wartofsky (eds.): *Methodology, Metaphysics and the History of Science.* In Memory of Benjamin Nelson. 1984 ISBN 90-277-1711-7
85. G. Tams: *The Logic of Categories.* Translated from Hungarian. Edited by R.S. Cohen. 1986 ISBN 90-277-1742-7
86. S.L. de C. Fernandes: *Foundations of Objective Knowledge.* The Relations of Popper's Theory of Knowledge to That of Kant. 1985 ISBN 90-277-1809-1
87. R.S. Cohen and T. Schnelle (eds.): *Cognition and Fact.* Materials on Ludwik Fleck. 1986 ISBN 90-277-1902-0
88. G. Freudenthal: *Atom and Individual in the Age of Newton.* On the Genesis of the Mechanistic World View. Translated from German. 1986 ISBN 90-277-1905-5
89. A. Donagan, A.N. Perovich Jr and M.V. Wedin (eds.): *Human Nature and Natural Knowledge.* Essays presented to Marjorie Grene on the Occasion of Her 75th Birthday. 1986 ISBN 90-277-1974-8
90. C. Mitcham and A. Hunning (eds.): *Philosophy and Technology II.* Information Technology and Computers in Theory and Practice. [*Also* Philosophy and Technology Series, Vol. 2] 1986 ISBN 90-277-1975-6
91. M. Grene and D. Nails (eds.): *Spinoza and the Sciences.* 1986 ISBN 90-277-1976-4
92. S.P. Turner: *The Search for a Methodology of Social Science.* Durkheim, Weber, and the 19th-Century Problem of Cause, Probability, and Action. 1986. ISBN 90-277-2067-3
93. I.C. Jarvie: *Thinking about Society.* Theory and Practice. 1986 ISBN 90-277-2068-1
94. E. Ullmann-Margalit (ed.): *The Kaleidoscope of Science.* The Israel Colloquium: Studies in History, Philosophy, and Sociology of Science, Vol. 1. 1986 ISBN 90-277-2158-0; Pb 90-277-2159-9
95. E. Ullmann-Margalit (ed.): *The Prism of Science.* The Israel Colloquium: Studies in History, Philosophy, and Sociology of Science, Vol. 2. 1986 ISBN 90-277-2160-2; Pb 90-277-2161-0
96. G. Mrkus: *Language and Production.* A Critique of the Paradigms. Translated from French. 1986 ISBN 90-277-2169-6
97. F. Amrine, F.J. Zucker and H. Wheeler (eds.): *Goethe and the Sciences: A Reappraisal.* 1987 ISBN 90-277-2265-X; Pb 90-277-2400-8
98. J.C. Pitt and M. Pera (eds.): *Rational Changes in Science.* Essays on Scientific Reasoning. Translated from Italian. 1987 ISBN 90-277-2417-2
99. O. Costa de Beauregard: *Time, the Physical Magnitude.* 1987 ISBN 90-277-2444-X
100. A. Shimony and D. Nails (eds.): *Naturalistic Epistemology.* A Symposium of Two Decades. 1987 ISBN 90-277-2337-0
101. N. Rotenstreich: *Time and Meaning in History.* 1987 ISBN 90-277-2467-9
102. D.B. Zilberman: *The Birth of Meaning in Hindu Thought.* Edited by R.S. Cohen. 1988 ISBN 90-277-2497-0
103. T.F. Glick (ed.): *The Comparative Reception of Relativity.* 1987 ISBN 90-277-2498-9
104. Z. Harris, M. Gottfried, T. Ryckman, P. Mattick Jr, A. Daladier, T.N. Harris and S. Harris: *The Form of Information in Science.* Analysis of an Immunology Sublanguage. With a Preface by Hilary Putnam. 1989 ISBN 90-277-2516-0
105. F. Burwick (ed.): *Approaches to Organic Form.* Permutations in Science and Culture. 1987 ISBN 90-277-2541-1

Boston Studies in the Philosophy of Science

106. M. Almsi: *The Philosophy of Appearances.* Translated from Hungarian. 1989
ISBN 90-277-2150-5
107. S. Hook, W.L. O'Neill and R. O'Toole (eds.): *Philosophy, History and Social Action.* Essays in Honor of Lewis Feuer. With an Autobiographical Essay by L. Feuer. 1988
ISBN 90-277-2644-2
108. I. Hronszky, M. FehÇr and B. Dajka: *Scientific Knowledge Socialized.* Selected Proceedings of the 5th Joint International Conference on the History and Philosophy of Science organized by the IUHPS (VeszprÇm, Hungary, 1984). 1988 ISBN 90-277-2284-6
109. P. Tillers and E.D. Green (eds.): *Probability and Inference in the Law of Evidence.* The Uses and Limits of Bayesianism. 1988 ISBN 90-277-2689-2
110. E. Ullmann-Margalit (ed.): *Science in Reflection.* The Israel Colloquium: Studies in History, Philosophy, and Sociology of Science, Vol. 3. 1988
ISBN 90-277-2712-0; Pb 90-277-2713-9
111. K. Gavroglu, Y. Goudaroulis and P. Nicolacopoulos (eds.): *Imre Lakatos and Theories of Scientific Change.* 1989 ISBN 90-277-2766-X
112. B. Glassner and J.D. Moreno (eds.): *The Qualitative-Quantitative Distinction in the Social Sciences.* 1989 ISBN 90-277-2829-1
113. K. Arens: *Structures of Knowing.* Psychologies of the 19th Century. 1989
ISBN 0-7923-0009-2
114. A. Janik: *Style, Politics and the Future of Philosophy.* 1989 ISBN 0-7923-0056-4
115. F. Amrine (ed.): *Literature and Science as Modes of Expression.* With an Introduction by S. Weininger. 1989 ISBN 0-7923-0133-1
116. J.R. Brown and J. Mittelstrass (eds.): *An Intimate Relation.* Studies in the History and Philosophy of Science. Presented to Robert E. Butts on His 60th Birthday. 1989
ISBN 0-7923-0169-2
117. F. D'Agostino and I.C. Jarvie (eds.): *Freedom and Rationality.* Essays in Honor of John Watkins. 1989 ISBN 0-7923-0264-8
118. D. Zolo: *Reflexive Epistemology.* The Philosophical Legacy of Otto Neurath. 1989
ISBN 0-7923-0320-2
119. M. Kearn, B.S. Philips and R.S. Cohen (eds.): *Georg Simmel and Contemporary Sociology.* 1989 ISBN 0-7923-0407-1
120. T.H. Levere and W.R. Shea (eds.): *Nature, Experiment and the Science.* Essays on Galileo and the Nature of Science. In Honour of Stillman Drake. 1989 ISBN 0-7923-0420-9
121. P. Nicolacopoulos (ed.): *Greek Studies in the Philosophy and History of Science.* 1990
ISBN 0-7923-0717-8
122. R. Cooke and D. Costantini (eds.): *Statistics in Science.* The Foundations of Statistical Methods in Biology, Physics and Economics. 1990 ISBN 0-7923-0797-6
123. P. Duhem: *The Origins of Statics.* Translated from French by G.F. Leneaux, V.N. Vagliente and G.H. Wagner. With an Introduction by S.L. Jaki. 1991 ISBN 0-7923-0898-0
124. H. Kamerlingh Onnes: *Through Measurement to Knowledge.* The Selected Papers, 1853-1926. Edited and with an Introduction by K. Gavroglu and Y. Goudaroulis. 1991
ISBN 0-7923-0825-5
125. M. Čapek: *The New Aspects of Time: Its Continuity and Novelties.* Selected Papers in the Philosophy of Science. 1991 ISBN 0-7923-0911-1
126. S. Unguru (ed.): *Physics, Cosmology and Astronomy, 1300–1700.* Tension and Accommodation. 1991 ISBN 0-7923-1022-5

Boston Studies in the Philosophy of Science

127. Z. Bechler: *Newton's Physics on the Conceptual Structure of the Scientific Revolution.* 1991 ISBN 0-7923-1054-3
128. É. Meyerson: *Explanation in the Sciences.* Translated from French by M-A. Siple and D.A. Siple. 1991 ISBN 0-7923-1129-9
129. A.I. Tauber (ed.): *Organism and the Origins of Self.* 1991 ISBN 0-7923-1185-X
130. F.J. Varela and J-P. Dupuy (eds.): *Understanding Origins.* Contemporary Views on the Origin of Life, Mind and Society. 1992 ISBN 0-7923-1251-1
131. G.L. Pandit: *Methodological Variance.* Essays in Epistemological Ontology and the Methodology of Science. 1991 ISBN 0-7923-1263-5
132. G. MunÇvar (ed.): *Beyond Reason.* Essays on the Philosophy of Paul Feyerabend. 1991 ISBN 0-7923-1272-4
133. T.E. Uebel (ed.): *Rediscovering the Forgotten Vienna Circle.* Austrian Studies on Otto Neurath and the Vienna Circle. Partly translated from German. 1991 ISBN 0-7923-1276-7
134. W.R. Woodward and R.S. Cohen (eds.): *World Views and Scientific Discipline Formation.* Science Studies in the [former] German Democratic Republic. Partly translated from German by W.R. Woodward. 1991 ISBN 0-7923-1286-4
135. P. Zambelli: *The Speculum Astronomiae and Its Enigma.* Astrology, Theology and Science in Albertus Magnus and His Contemporaries. 1992 ISBN 0-7923-1380-1
136. P. Petitjean, C. Jami and A.M. Moulin (eds.): *Science and Empires.* Historical Studies about Scientific Development and European Expansion. ISBN 0-7923-1518-9
137. W.A. Wallace: *Galileo's Logic of Discovery and Proof.* The Background, Content, and Use of His Appropriated Treatises on Aristotle's *Posterior Analytics.* 1992 ISBN 0-7923-1577-4
138. W.A. Wallace: *Galileo's Logical Treatises.* A Translation, with Notes and Commentary, of His Appropriated Latin Questions on Aristotle's *Posterior Analytics.* 1992 ISBN 0-7923-1578-2
Set (137 + 138) ISBN 0-7923-1579-0
139. M.J. Nye, J.L. Richards and R.H. Stuewer (eds.): *The Invention of Physical Science.* Intersections of Mathematics, Theology and Natural Philosophy since the Seventeenth Century. Essays in Honor of Erwin N. Hiebert. 1992 ISBN 0-7923-1753-X
140. G. Corsi, M.L. dalla Chiara and G.C. Ghirardi (eds.): *Bridging the Gap: Philosophy, Mathematics and Physics.* Lectures on the Foundations of Science. 1992 ISBN 0-7923-1761-0
141. C.-H. Lin and D. Fu (eds.): *Philosophy and Conceptual History of Science in Taiwan.* 1992 ISBN 0-7923-1766-1
142. S. Sarkar (ed.): *The Founders of Evolutionary Genetics.* A Centenary Reappraisal. 1992 ISBN 0-7923-1777-7
143. J. Blackmore (ed.): *Ernst Mach – A Deeper Look.* Documents and New Perspectives. 1992 ISBN 0-7923-1853-6
144. P. Kroes and M. Bakker (eds.): *Technological Development and Science in the Industrial Age.* New Perspectives on the Science–Technology Relationship. 1992 ISBN 0-7923-1898-6
145. S. Amsterdamski: *Between History and Method.* Disputes about the Rationality of Science. 1992 ISBN 0-7923-1941-9
146. E. Ullmann-Margalit (ed.): *The Scientific Enterprise.* The Bar-Hillel Colloquium: Studies in History, Philosophy, and Sociology of Science, Volume 4. 1992 ISBN 0-7923-1992-3
147. L. Embree (ed.): *Metaarchaeology.* Reflections by Archaeologists and Philosophers. 1992 ISBN 0-7923-2023-9
148. S. French and H. Kamminga (eds.): *Correspondence, Invariance and Heuristics.* Essays in Honour of Heinz Post. 1993 ISBN 0-7923-2085-9
149. M. Bunzl: *The Context of Explanation.* 1993 ISBN 0-7923-2153-7

Boston Studies in the Philosophy of Science

150. I.B. Cohen (ed.): *The Natural Sciences and the Social Sciences.* Some Critical and Historical Perspectives. 1994 ISBN 0-7923-2223-1
151. K. Gavroglu, Y. Christianidis and E. Nicolaidis (eds.): *Trends in the Historiography of Science.* 1994 ISBN 0-7923-2255-X
152. S. Poggi and M. Bossi (eds.): *Romanticism in Science.* Science in Europe, 1790–1840. 1994 ISBN 0-7923-2336-X
153. J. Faye and H.J. Folse (eds.): *Niels Bohr and Contemporary Philosophy.* 1994 ISBN 0-7923-2378-5
154. C.C. Gould and R.S. Cohen (eds.): *Artifacts, Representations, and Social Practice.* Essays for Marx W. Wartofsky. 1994 ISBN 0-7923-2481-1
155. R.E. Butts: *Historical Pragmatics.* Philosophical Essays. 1993 ISBN 0-7923-2498-6
156. R. Rashed: *The Development of Arabic Mathematics: Between Arithmetic and Algebra.* Translated from French by A.F.W. Armstrong. 1994 ISBN 0-7923-2565-6
157. I. Szumilewicz-Lachman (ed.): *Zygmunt Zawirski: His Life and Work.* With Selected Writings on Time, Logic and the Methodology of Science. Translations by Feliks Lachman. Ed. by R.S. Cohen, with the assistance of B. Bergo. 1994 ISBN 0-7923-2566-4
158. S.N. Haq: *Names, Natures and Things.* The Alchemist Jābir ibn Ḥayyān and His *Kitāb al-Aḥjār* (Book of Stones). 1994 ISBN 0-7923-2587-7
159. P. Plaass: *Kant's Theory of Natural Science.* Translation, Analytic Introduction and Commentary by Alfred E. and Maria G. Miller. 1994 ISBN 0-7923-2750-0
160. J. Misiek (ed.): *The Problem of Rationality in Science and its Philosophy.* On Popper vs. Polanyi. The Polish Conferences 1988–89. 1995 ISBN 0-7923-2925-2
161. I.C. Jarvie and N. Laor (eds.): *Critical Rationalism, Metaphysics and Science.* Essays for Joseph Agassi, Volume I. 1995 ISBN 0-7923-2960-0
162. I.C. Jarvie and N. Laor (eds.): *Critical Rationalism, the Social Sciences and the Humanities.* Essays for Joseph Agassi, Volume II. 1995 ISBN 0-7923-2961-9
Set (161–162) ISBN 0-7923-2962-7
163. K. Gavroglu, J. Stachel and M.W. Wartofsky (eds.): *Physics, Philosophy, and the Scientific Community.* Essays in the Philosophy and History of the Natural Sciences and Mathematics. In Honor of Robert S. Cohen. 1995 ISBN 0-7923-2988-0
164. K. Gavroglu, J. Stachel and M.W. Wartofsky (eds.): *Science, Politics and Social Practice.* Essays on Marxism and Science, Philosophy of Culture and the Social Sciences. In Honor of Robert S. Cohen. 1995 ISBN 0-7923-2989-9
165. K. Gavroglu, J. Stachel and M.W. Wartofsky (eds.): *Science, Mind and Art.* Essays on Science and the Humanistic Understanding in Art, Epistemology, Religion and Ethics. Essays in Honor of Robert S. Cohen. 1995 ISBN 0-7923-2990-2
Set (163–165) ISBN 0-7923-2991-0
166. K.H. Wolff: *Transformation in the Writing.* A Case of Surrender-and-Catch. 1995 ISBN 0-7923-3178-8
167. A.J. Kox and D.M. Siegel (eds.): *No Truth Except in the Details.* Essays in Honor of Martin J. Klein. 1995 ISBN 0-7923-3195-8
168. J. Blackmore: *Ludwig Boltzmann, His Later Life and Philosophy, 1900–1906.* Book One: A Documentary History. 1995 ISBN 0-7923-3231-8
169. R.S. Cohen, R. Hilpinen and R. Qiu (eds.): *Realism and Anti-Realism in the Philosophy of Science.* Beijing International Conference, 1992. 1996 ISBN 0-7923-3233-4
170. I. Kuçuradi and R.S. Cohen (eds.): *The Concept of Knowledge.* The Ankara Seminar. 1995 ISBN 0-7923-3241-5

Boston Studies in the Philosophy of Science

171. M.A. Grodin (ed.): *Meta Medical Ethics*: The Philosophical Foundations of Bioethics. 1995 ISBN 0-7923-3344-6
172. S. Ramirez and R.S. Cohen (eds.): *Mexican Studies in the History and Philosophy of Science.* 1995 ISBN 0-7923-3462-0
173. C. Dilworth: *The Metaphysics of Science.* An Account of Modern Science in Terms of Principles, Laws and Theories. 1995 ISBN 0-7923-3693-3
174. J. Blackmore: *Ludwig Boltzmann, His Later Life and Philosophy, 1900–1906* Book Two: The Philosopher. 1995 ISBN 0-7923-3464-7
175. P. Damerow: *Abstraction and Representation.* Essays on the Cultural Evolution of Thinking. 1996 ISBN 0-7923-3816-2
176. M.S. Macrakis: *Scarcity's Ways: The Origins of Capital.* A Critical Essay on Thermodynamics, Statistical Mechanics and Economics. 1997 ISBN 0-7923-4760-9
177. M. Marion and R.S. Cohen (eds.): *QuÇbec Studies in the Philosophy of Science.* Part I: Logic, Mathematics, Physics and History of Science. Essays in Honor of Hugues Leblanc. 1995 ISBN 0-7923-3559-7
178. M. Marion and R.S. Cohen (eds.): *QuÇbec Studies in the Philosophy of Science.* Part II: Biology, Psychology, Cognitive Science and Economics. Essays in Honor of Hugues Leblanc. 1996 ISBN 0-7923-3560-0
     Set (177–178) ISBN 0-7923-3561-9
179. Fan Dainian and R.S. Cohen (eds.): *Chinese Studies in the History and Philosophy of Science and Technology.* 1996 ISBN 0-7923-3463-9
180. P. Forman and J.M. Snchez-Ron (eds.): *National Military Establishments and the Advancement of Science and Technology.* Studies in 20th Century History. 1996 ISBN 0-7923-3541-4
181. E.J. Post: *Quantum Reprogramming.* Ensembles and Single Systems: A Two-Tier Approach to Quantum Mechanics. 1995 ISBN 0-7923-3565-1
182. A.I. Tauber (ed.): *The Elusive Synthesis: Aesthetics and Science.* 1996 ISBN 0-7923-3904-5
183. S. Sarkar (ed.): *The Philosophy and History of Molecular Biology: New Perspectives.* 1996 ISBN 0-7923-3947-9
184. J.T. Cushing, A. Fine and S. Goldstein (eds.): *Bohmian Mechanics and Quantum Theory: An Appraisal.* 1996 ISBN 0-7923-4028-0
185. K. Michalski: *Logic and Time.* An Essay on Husserl's Theory of Meaning. 1996 ISBN 0-7923-4082-5
186. G. MunÇvar (ed.): *Spanish Studies in the Philosophy of Science.* 1996 ISBN 0-7923-4147-3
187. G. Schubring (ed.): *Hermann G~nther Graßmann (1809–1877): Visionary Mathematician, Scientist and Neohumanist Scholar.* Papers from a Sesquicentennial Conference. 1996 ISBN 0-7923-4261-5
188. M. Bitbol: *Schrdinger's Philosophy of Quantum Mechanics.* 1996 ISBN 0-7923-4266-6
189. J. Faye, U. Scheffler and M. Urchs (eds.): *Perspectives on Time.* 1997 ISBN 0-7923-4330-1
190. K. Lehrer and J.C. Marek (eds.): *Austrian Philosophy Past and Present.* Essays in Honor of Rudolf Haller. 1996 ISBN 0-7923-4347-6
191. J.L. Lagrange: *Analytical Mechanics.* Translated and edited by Auguste Boissonade and Victor N. Vagliente. Translated from the *MÇcanique Analytique, novelle Çdition* of 1811. 1997 ISBN 0-7923-4349-2
192. D. Ginev and R.S. Cohen (eds.): *Issues and Images in the Philosophy of Science.* Scientific and Philosophical Essays in Honour of Azarya Polikarov. 1997 ISBN 0-7923-4444-8

Boston Studies in the Philosophy of Science

193. R.S. Cohen, M. Horne and J. Stachel (eds.): *Experimental Metaphysics.* Quantum Mechanical Studies for Abner Shimony, Volume One. 1997 ISBN 0-7923-4452-9
194. R.S. Cohen, M. Horne and J. Stachel (eds.): *Potentiality, Entanglement and Passion-at-a-Distance.* Quantum Mechanical Studies for Abner Shimony, Volume Two. 1997 ISBN 0-7923-4453-7; Set 0-7923-4454-5
195. R.S. Cohen and A.I. Tauber (eds.): *Philosophies of Nature: The Human Dimension.* 1997 ISBN 0-7923-4579-7
196. M. Otte and M. Panza (eds.): *Analysis and Synthesis in Mathematics.* History and Philosophy. 1997 ISBN 0-7923-4570-3
197. A. Denkel: *The Natural Background of Meaning.* 1999 ISBN 0-7923-5331-5
198. D. Baird, R.I.G. Hughes and A. Nordmann (eds.): *Heinrich Hertz: Classical Physicist, Modern Philosopher.* 1999 ISBN 0-7923-4653-X
199. A. Franklin: *Can That be Right?* Essays on Experiment, Evidence, and Science. 1999 ISBN 0-7923-5464-8
200. Reserved
201. Reserved
202. Reserved
203. B. Babich and R.S. Cohen (eds.): *Nietzsche, Theories of Knowledge, and Critical Theory.* Nietzsche and the Sciences I. 1999 ISBN 0-7923-5742-6
204. B. Babich and R.S. Cohen (eds.): *Nietzsche, Epistemology, and Philosophy of Science.* Nietzsche and the Science II. 1999 ISBN 0-7923-5778-7
205. R. Hooykaas; H.F. Cohen, J.C. Boudri and V. Mackay (eds.): *Fact, Faith and Fiction in the Development of Science.* The Gifford Lectures given in the University of St Andrews 1976. 1999 ISBN 0-7923-5774-4
206. M. Fehér, O. Kiss and L. Ropolyi (eds.): *Hermeneutics and Science.* 1999 ISBN 0-7923-5798-1
207. R.M. MacLeod (ed.): *Science and the Pacific War.* Science and Survival in the Pacific, 1939-1945. 1999 ISBN 0-7923-5851-1
208. I. Hanzel: *The Concept of Scientific Law in the Philosophy of Science and Epistemology.* A Study of Theoretical Reason. 1999 ISBN 0-7923-5852-X
209. G. Helm; R.J. Deltete (ed./transl.): *The Historical Development of Energetics.* 1999 ISBN 0-7923-5874-0
210. A. Orenstein and P. Kotatko (eds.): *Knowledge, Language and Logic.* Questions for Quine. 1999 ISBN 0-7923-5986-0

Also of interest:
R.S. Cohen and M.W. Wartofsky (eds.): *A Portrait of Twenty-Five Years Boston Colloquia for the Philosophy of Science, 1960-1985.* 1985 ISBN Pb 90-277-1971-3

Previous volumes are still available.

KLUWER ACADEMIC PUBLISHERS – DORDRECHT / BOSTON / LONDON

MIX
Papier aus verantwortungsvollen Quellen
Paper from responsible sources
FSC® C105338

If you have any concerns about our products,
you can contact us on
ProductSafety@springernature.com

In case Publisher is established outside the EU,
the EU authorized representative is:
Springer Nature Customer Service Center GmbH
Europaplatz 3, 69115 Heidelberg, Germany

Printed by Libri Plureos GmbH
in Hamburg, Germany